PROTEIN BIOCHEMISTRY, SYNTHESIS, STRUCTURE AND CELLULAR FUNCTIONS

TYROSINE AND ASPARTIC ACID: PROPERTIES, SOURCES AND HEALTH BENEFITS

PROTEIN BIOCHEMISTRY, SYNTHESIS, STRUCTURE AND CELLULAR FUNCTIONS

Additional books in this series can be found on Nova's website under the Series tab.

Additional E-books in this series can be found on Nova's website under the E-book tab.

PROTEIN BIOCHEMISTRY, SYNTHESIS, STRUCTURE AND CELLULAR FUNCTIONS

TYROSINE AND ASPARTIC ACID: PROPERTIES, SOURCES AND HEALTH BENEFITS

JANICE E. JONES
AND
DANIELLE M. MORANO
EDITORS

Nova Science Publishers, Inc.
New York

LIBRARY OF CONGRESS CATALOGING-IN-PUBLICATION DATA

Tyrosine and aspartic acid : properties, sources, and health benefits / editors, Janice E. Jones and Danielle M. Morano.
p. ; cm.
Includes bibliographical references and index.
ISBN 978-1-62100-752-4 (hardcover)
I. Jones, Janice E. II. Morano, Danielle M.
[DNLM: 1. Tyrosine--physiology. 2. Aspartic Acid--physiology. 3. Aspartic Acid--therapeutic use. 4. Tyrosine--therapeutic use. QU 60]

616.7'227--dc23

2011038573

Published by Nova Science Publishers, Inc. † New York

Contents

PREFACE

This book presents current research in the study of the properties, sources and health benefits of the amino acids tyrosine and aspartic acid. Topics discussed include tyrosine kinases and rheumatoid arthritis; aspartic acid racemization in the study of ancient DNA in bones; the molecular dynamics simulation for stereoinversion of aspartic acid residues; the physiological role of RTK transactivation in the central nervous system and neuropsychiatric diseases and tyrosines in biological electron transfer.

Chapter 1- Rheumatoid arthritis (RA) is an autoimmune disease characterized by the accumulation and proliferation of inflammatory cells in the synovial (joint) lining, resulting in the formation of pannus tissue, which invades and destroys adjacent cartilage and bone. In RA macrophages, B cells, mast cells, fibroblast-like synoviocytes (FLSs) and $CD4^+$ T lymphocytes become activated and contribute to synovial inflammation and joint destruction. It has been showed that different tyrosine kinases participate in the activation of those cells having important participation in the physiopathology of RA. Therefore, the tyrosine kinases inhibitors could be the next step in the treatment of patients with RA. This review focuses on recent advances on the role of tyrosine kinases and their inhibitors in the physiopathology of RA.

Chapter 2- Specific amino acid side chains are crucial to the stability of proteins structure and in the catalysis of enzymatic reactions. An important role is granted to Tyr residues and Tyr derivatives originating from posttranslational modifications. An example for such a modified Tyr is found in cytochrome *c* oxidase, where a covalent linkage between a Tyr and a His was found. The role of this linkage for the catalytic mechanism was explored by a number of research groups, several of the studies including model compound studies. Electrochemical and spectroscopic studies gave evidence to the influence of this linkage on the electron transfer ability of the Tyr and its function within the catalytic mechanism.

Chapter 3- The ability of memory CD4+ T cells (CD45RO) to better replicate HIV-1 in comparison to naïve CD4+ T cells (CD45RA) have been well established. In a Jurkat cell system, we had previously suggested that the nuclear factor of activated T cells (NFAT) was a key element for the higher HIV-1 LTR activation level and the more intense HIV-1 replication observed in Jurkat cells expressing the CD45RO isoform (as compared to CD45RA-expressing cells). To further investigate on the step of the HIV-1 replicative cycle required for preferential replication in memory CD4+ T cells, primary CD4+ T lymphocytes were sorted in naïve and memory CD4+ T cell subsets using antibody-coupled magnetic beads and subsequently infected. We first confirmed that wild-type and pseudotyped luciferase-expressing HIV-1 virions showed stronger replication in memory CD4+ T lymphocytes following cellular activation. Early measurement of intracellular p24 after contact of T lymphocytes with HIV-1 virions and experiments using VSV-G-pseudotyped HIV-1 virions indicated that viral entry was comparable in both memory and naïve CD4+ T cells. Through real time PCR analysis, we demonstrated that the content in viral DNA (total or integrated) was higher in the memory CD4+ T cell population upon activation. The involvement of NFAT in this process was strongly suggested by Electrophoretic Mobility Shift Assays and infection experiments conducted in the presence of the NFAT inhibitor FK506. Our results thereby argues that reverse transcription is an important step restricting HIV-1 replication to memory CD4+ T cells and that NFAT would be an important regulator of this process.

Chapter 4- Only a small proportion of old specimens contain DNA that can be amplified by PCR: in the study of ancient DNA, reliable methods for the identification of the large fraction of samples that are so badly preserved that there is no need to attempt DNA extraction, especially in bones and teeth, are very useful. In particular, the extent of racemization of some amino acids has proved to be a very useful proxy for DNA preservation.

Almost all amino acids can occur in two optical isomers, termed L and D. Most naturally occurring proteins contain only L-amino acids, although there are some exceptional peptides which are biologically synthesized using D-form amino acids. Metabolically active tissues contain specialized enzymes known as racemases which maintain a disequilibrium in our cells of only the L isomers. Once an organism dies and its enzymes cease to function, amino acids undergo racemization, that is, the interconversion of L amino acids to D amino acids. Racemization is affected by some of the factors that affect degradation of DNA. In particular, the rate of aspartic acid racemization is nearly identical to the rate of DNA depurination: in general, no DNA sequences of sufficient length can be extracted from samples in which the D/L Asp ratio is higher than 0.1.

In this chapter we review the major published methods in this field, especially referring to sample preparation, protein hydrolysis and D/L Asp ratio determination in fossil bones.

Chapter 5- Recently, stereoinversions of aspartic acid residues were observed in some proteins and peptides. Because D-aspartic acids were found in crystallin from cataract patients and in β-amyloid from Alzheimer's dementia patients, it is suspected that the stereoinversion of aspartic acid residues is relevant in age-related diseases. The stereoinversion of aspartic acid residues should induce structural changes in their associated proteins and peptides, and therefore, structural investigations of these modified proteins and peptides are essential. Furthermore, proteins that incorporate D-aspartic acid residues (e.g., crystallin and β-amyloid) tend to aggregate, making experimental investigations of their atomic-level structures difficult. Thus, structural bioinformatics methods play an important role in studies of D-aspartic acid residues. We employed the molecular dynamics (MD) method, which is one of the most useful structural bioinformatics procedures, to investigate the stereoinversion-related properties of proteins. In this chapter, an evaluation of various parameters used in MD simulations was performed. Because parameters were developed for L-aspartic acid, their usefulness for D-aspartic acids was unknown. We computationally constructed peptides including D-aspartic acid residues and performed MD simulations with these peptides. The results suggest that the parameters developed for L-aspartic acid are also appropriate for D-aspartic acid residues.

Chapter 6- Over the past years, it has been demonstrated that amino acids and some neurotransmitters –serotonin and dopamine– are responsible for the most of synaptic communication in our brain. Along the time, research has been basically focused to L-enantiomers of amino acids. L-Amino acids build blocks of proteins; each protein is composed of a chain made up of amino acids in a unique sequence. Prior to the 80´s decade, L-amino acids were considered as vital for different functions in living organisms. However, due to the recent advances of analytical techniques and methods, the presence of D-amino acids has been so far recognized in the biological world with essential physiological functions. Besides these advances in analytical techniques, it has also allowed a more accurate measurement of small amounts of D-amino acids among much larger pools of L-amino acids in biological fluids and tissues (Friedman and Levin, 2011; Ilisz et al. 2010; Kirschner and Green, 2009). For example, D-Aspartic acid (D-Asp) –considered as one of the most studied D-amino acids– is an endogenous amino acid found in nervous and endocrine tissues of invertebrates and vertebrates. D-Asp promotes the proteins synthesis in the development of the nervous system and regulates the hormonal release in the endocrine glands. D-Asp –which acts as a neurotransmitter or neuromodulator– is one of the major excitatory amino acids in the brain. Consequently, for a better understanding of its biological importance, it is crucial to study their occurrence and metabolism in living systems. In this work, the study will be focused in the role of D-Asp in central nervous system.

Chapter 7- Amide–iminol tautomerization is an analogue of the well-known keto–enol tautomerization. In this chapter, we show, based on quantum-chemical calculations for a model compound, that iminolization of the C-terminal peptide groups of aspartic acid (Asp) residues may actually occur in peptides and proteins with assistance by two catalytic water molecules. Moreover, it is shown that the iminol tautomer can undergo a two-water-assisted cyclization with relative ease to form a *gem*-diol, which is the precursor of the well-known succinimide imtermediates of isomerization and racemization reactions of Asp residues. These findings indicate that the long-established succinimide-mediated mechanism of the Asp residue reactions may have to be somewhat modified. We also point out that the iminolization mechanism found in the present study may be responsible for the acid-catalyzed hydrogen exchange of the C-terminal peptide groups of Asp residues.

Chapter 8- G protein-coupled receptors (GPCRs) generate signals that promote gene transcription through the 'transactivation' of receptor tyrosine kinases (RTKs) such as the receptors for epidermal growth factor, fibroblast growth factor, platelet-derived growth factor (PDGF), and neurotrophins. RTKs and downstream signaling pathways activated in response to GPCR-mediated stimulation elicit phenotypic responses. Transactivation is essential for the specificity and fidelity of signal transduction and biological responses and is particularly important in the central nervous system (CNS), where RTKs mediate key events in the development and maintenance of neuronal circuits. Recent evidence indicates that RTK transactivation may be involved in the pathophysiology of neuropsychiatric diseases. Dopamine, a major transmitter and neuromodulator in the CNS, is implicated in psychiatric diseases. A GPCR, the D4 dopamine receptor, inhibits 1) excitatory transmission mediated by the N-methyl-D-aspartate (NMDA) glutamate receptor subtype and 2) synaptic transmission through transactivation of the PDGF receptor. The demonstration of a physiological role for RTK transactivation in the CNS provides opportunities for understanding how aberrant dopamine signaling might contribute to neuropsychiatric diseases and for producing novel treatment strategies for these diseases.

Chapter 9- L-Aspartate (L-Asp) functions as a neurotransmitter to stimulate the N-methyl-D-aspartate (NMDA) receptor (NMDA-R), one of the ionotropic L-glutamate receptors, even though its binding capacity for NMDA-R is weaker than L-glutamate. The amino acid L-Asp, as well as its enantiomer D-aspartate (D-Asp), occurs in the central nervous system of various species including chickens, pigeons, rats, mice and humans. D-Asp, synthesized from L-Asp by aspartate racemase, can also directly stimulate the NMDA-R. Furthermore, D-Asp may indirectly stimulate NMDA-R because it is a substrate for endogenous NMDA. Central injection of L-Asp and L-asparagine derived from L-Asp via transamidation decreased social separation-induced stress responses. In addition, central injection of NMDA can attenuate the stress responses while some reports indicate that stimulation of the NMDA-R has negative impacts on stress responses. Central L-Asp may act on stress responses not only directly but also via its metabolites, and the effects may depend on the type of stressors and/or brain regions.

In: Tyrosine and Aspartic Acid
Editors: J. E. Jones and D. M. Morano
ISBN: 978-1-62100-752-4

Chapter 1

TYROSINE KINASES AND RHEUMATOID ARTHRITIS

Antonio G. Tristano*
Centro Medico Docente La Trinidad, Caracas Venezuela, and Centro Médico Carpetana Madrid Spain

ABSTRACT

Rheumatoid arthritis (RA) is an autoimmune disease characterized by the accumulation and proliferation of inflammatory cells in the synovial (joint) lining, resulting in the formation of pannus tissue, which invades and destroys adjacent cartilage and bone. In RA macrophages, B cells, mast cells, fibroblast-like synoviocytes (FLSs) and $CD4^+$ T lymphocytes become activated and contribute to synovial inflammation and joint destruction. It has been showed that different tyrosine kinases participate in the activation of those cells having important participation in the physiopathology of RA. Therefore, the tyrosine kinases inhibitors could be the next step in the treatment of patients with RA. This review focuses on recent advances on the role of tyrosine kinases and their inhibitors in the physiopathology of RA.

Keywords: Rheumatoid arthritis; tyrosine kinases; tyrosine kinases inhibitors; dasatinib; imatinib: tofacitinib

INTRODUCTION

Rheumatoid arthritis (RA) is an autoimmune disease affecting 0.5%-1% of the world population. RA is characterized by the accumulation and proliferation of inflammatory cells in the synovial (joint) lining, resulting in the formation of pannus tissue, which invades and destroys adjacent cartilage and bone [1]. Although the aetiology of RA remains unknown, macrophages, B cells, mast cells, fibroblast-like synoviocytes (FLSs) and $CD4^+$ T lymphocytes become activated and contribute to synovial inflammation and joint destruction [2].

* E-mail: antoniotristano@yahoo.com, Mail address: 5150 NW 99th Ave. Miami Florida 33178 USA.

In RA: a) macrophages infiltrate the synovium and secrete tumour necrosis factor-alpha (TNF-α) and other proinflammatory cytokines that potentiate inflammation [3,4]; b) mast cell activation through c-Kit (receptor tyrosine kinase critical for mast cell development and activation) results in the release of mediators that contribute to the inflammatory and degradative processes, including histamine, heparin, neutral proteases, and TNF-α [5-8, 9,10]; c) fibroblasts express PDGFR (platelet-derived growth factor receptor) and proliferate in response to a variety of PDGF ligands. Both PDGFR and its ligands are over expressed in RA synovial tissue, and PDGF (platelet-derived growth factor) is a potent stimulant of synovial hyperplasia [11-13].

Furthermore, bone and cartilage destruction is one of the most serious complications in RA patients, which can result in progressive morbidity and significant disability. The cells involved in this complication have been extensively studied and although the mechanisms of cartilage destruction in RA were well described, the specific mechanisms responsible for bone erosion in this disease have only recently been studied. The studies of bone erosion in patients with RA and in animal models of inflammatory arthritis such as collagen-induced arthritis (CIA) have provided powerful evidence that bone-resorbing osteoclasts play an important role in the structural joint damage involved in inflammatory arthritis [14]. Osteoclast precursors express receptor activator of NF-κB (RANK) and differentiate osteoclasts in the presence of receptor activator of NF-κB ligand (RANKL) and M-CSF [15-17]. RA synovial tissues produce M-CSF, RANKL, and various cytokines that could increase osteoclast formation or activity, including interleukin 1-alpha (IL-1α and -beta (IL-1β), TNF-α, IL-6, IL-11, and IL- 17 [18-21].

There are evidences that tyrosine kinases inhibitors (TKIs) have the ability of inhibiting mast cell c-Kit, FLS PDGFR, macrophage c-Fms, the phosphorylation of c-*fms*, a receptor for macrophage colony-stimulating factor (M-CSF) [22] that is an essential factor for osteoclast formation [23-26], and inhibit TCR-mediated signal transduction, cellular proliferation, cytokine production, and in vivo T cell responses [27]. All these factors have important participation in the physiopathology of RA; therefore the TKIs could be the next step in the treatment of patients with RA. Consistent with these *in vitro* evidences, recent cases reports describe severe RA patients treated with TKIs, who show marked improvement in the joint pain and disease activity [28-29] and animal studies where the TKIs treat, prevent, and inhibit osteoclastogenesis and joint destruction in CIA [30-32].

Tyrosine kinases represent a diverse and rapidly expanding superfamily of proteins, including both transmembrane receptor tyrosine kinases (RTK) and soluble cytoplasmic enzymes also known as nonreceptor tyrosine kinases (NRTK). Tyrosine kinases modulate a wide variety of cellular events, including differentiation, growth, metabolism and apoptosis [33-37]. This is a review about the participation of tyrosine kinases in the physiopathology of RA and the possible therapeutic effects of TKIs. Specifically, this systemic review will cover some concepts about the tyrosine kinases and its inhibitors. Then it is going to be discussed the tyrosine kinases involved in RA such as the PDGFR in fibroblast, the c-kit in mast cells, the c-Fms and VEGFR in macrophage, the c-fms and c-Src in osteoclast, the TCR signaling in T cells and others kinases. Finally, the clinical and animal studies published about TKIs in RA will be discussed.

TYROSINE KINASES AND TYROSINE KINASES INHIBITORS

Tyrosine kinases modulate a wide variety of cellular events, including differentiation, growth, metabolism and apoptosis [33-37]. The human genome encodes for almost 600 proteins kinases. Approximately 100-200 proteins kinases are present in each cell, located upstream and downstream of epidemiologically relevant oncogenes or tumour suppressor genes [38]. Tyrosine kinases represent a diverse and rapidly expanding superfamily of proteins, including both transmembrane receptor tyrosine kinases (RTK) and soluble cytoplasmic enzymes also known as nonreceptor tyrosine kinases (NRTK). Common features of all tyrosine kinases include a separate domain for substrate binding, ATP binding, and catalysis [39]. The latter domain promotes the transfer of the terminal phosphoryl group from ATP to a tyrosine amino group acceptor in a substrate.

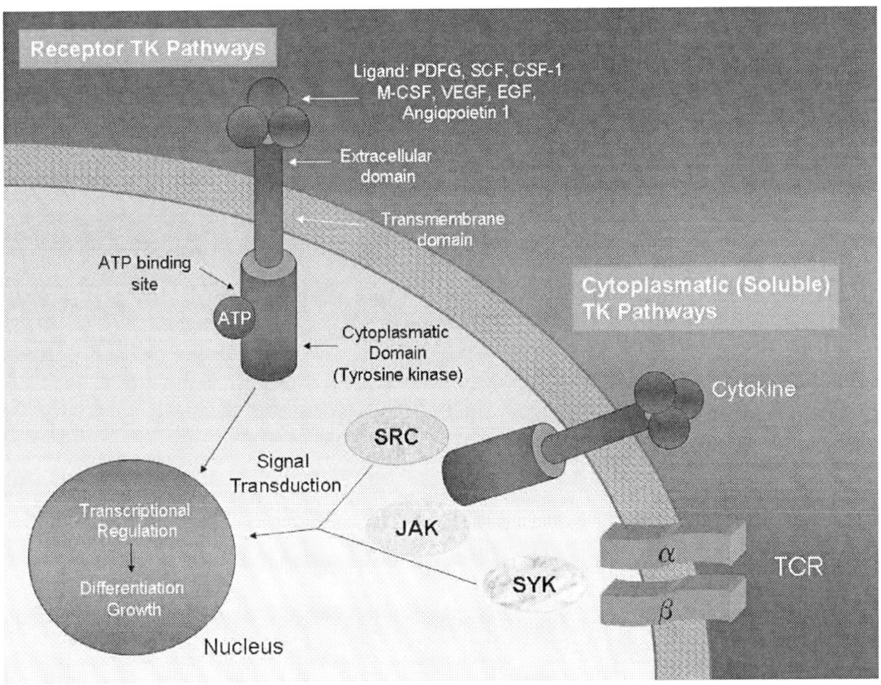

Figure 1. Schematic representations of the different tyrosine kinases signaling pathways involved in RA. From Tristano AG [167].

Phosphorylation of tyrosine residues in target proteins is essential for maintaining cellular homeostasis, yet this post-translational modification also provides the means by which a number of cellular oncogenes deregulate various signalling pathways and induce transformation [35]. Activation of the tyrosine kinases domain of either class of tyrosine kinases enzymes results in interaction of the protein with other signal transducing molecules and propagation of the signal along a specific signal transduction pathway (Figure 1)[33-37,40-44]. Activation of transmembrane tyrosine kinases is typically initiated by binding a ligand (e.g., hormone or growth factor) to a specific site within the extracellular domain of the receptor. Upon ligand binding, these receptors commonly undergo dimerization, resulting in autophosphorylation of tyrosine residues within the cytoplasmic domain [45,46]. These phosphorylation events activate the kinase, thereby increasing its intrinsic tyrosine kinases activity, and produce new binding sites for intracellular adapter molecules that bring signal transduction molecules into close proximity. Receptors that lack tyrosine kinases activity but harbor sites for tyrosine phosphorylation (often catalyzed by the soluble cytoplasmic tyrosine kinases enzymes) activate identical or similar enzyme cascades via the association of their phosphotyrosine residues with adapter molecules [43,47]. Significant overlap exists between these general signalling pathways, including the activation of the mitogen-activated/extracellular signal-regulated protein kinase (MAPK or ERK) superfamily [36,44].

There are different tyrosine kinases involved in the physiopathology of RA (Table 1, Figure 1), putting them as a target in the treatment of patients with RA. With the advance in molecular and cell biology technologies, an increased number of tyrosine kinases have been identified and as a consequence of that, the molecular targets have also greatly increased. With the approvals of the monoclonal antibody herceptin against breast cancer, the Bcr-Abl inhibitors imatinib and dasatinib for chronic myelogenous leukemia (CML), and the epidermal growth factor receptor inhibitor gefitinib, the first TKIs have entered the market [38].

Table 1. Classification and function of tyrosine kinases (TKs) involved in RA

		Transmembrane receptors	
TKs	Ligands	Effects in RA	Others diseases
PDGFR	PDFG	activation, transformation	gliomas
		and proliferation of FLSs	
SCF-R/KIT	SCF	increases the survival of MCs	GIST, SCLC, glioma
		and stimulates their proliferation	
CSF1- R/FMS	CSF-1	stimulating the survival,	leukemia
		proliferation and differentiation of	
		mononuclear phagocytes	
	M-CSF	osteoclastogenesis	
VEGF-R/FLT	VEGF	angiogenesis, inflammation	soild tomours, AML
ErbB-2	EGF	hyperplasia of rheumatoid synovialcells	Breast, lung, prostate
Tie2	Angiopoietin 1	Angiogenesi	
		Nonreceptors TKs	
TK Family	TKs	Effects in RA	Others diseases
SRC	c-Src, Lck	proliferation of activated synoviocytes	Solid tumours
	Fyn	elevated production of cytokines	
		osteoclastic bone resorption	
JAK	jak1, jak2	cellular activation, proliferation,	leukemia
	tyk2		
SYK	zap, syk	production of inflammatory cytokine	B and T cell
		and MMP production by FLS	(autoimmunity)

AML: acute myeloid leukaemia; FLS: fibroblast; GIST: gastrointestinal stromal tumours; MMP: metaloproteinases; RA: rheumatoid arthritis; SCLC: small cell lung cancer. From Tristano AG [167].

TYROSINE KINASES AND RA

Fibroblast and PDGFR

Synovial fibroblasts are activated and transformed into aggressive, matrix-degrading cells in patients with RA, so one could describe them as tumour-like, but in contrast to many tumours, RA synovial fibroblasts or fibroblast-like synoviocytes (FLSs) are not autonomous, instead responding to different inflammatory signals as cytokines [48,49]. However, RA-FLSs have been shown to accumulate a number of mutations similar to those seen in malignancy, including mutations in tumour suppressor genes [50]. These mutations may facilitate additional RA-FLSs proliferation. In this regard, the investigation of the mode of proliferation revealed an upregulation of the metastatic lymph node MLN51 gene (originally identified in breast cancer) in hyperactive RA-FLSs [51]. On the other hand, cultured FLSs cells from RA patients exhibit anchorage-independent growth and a loss of contact inhibition, especially in the presence of platelet-derived growth factor (PDGF) [52], as well as oncogene activation [53], monoclonal or oligoclonal cell expansion [54] and, most importantly, cartilage invasion in severe combined immunodeficient mice [55]. RA-FLSs have also the capacity to produce and secrete proinflammatory mediators, including cytokines, growth factors and lipid mediators of inflammation.

The FLSs growth factors secreted, including PDGF, fibroblast growth factor, vascular endothelial growth factor, epidermal growth factor, granulocyte-macrophage colony stimulating factor (GM-CSF), granulocyte colony stimulating factor (G-CSF), and TGF-β, all contribute to neoangiogenesis [56,57]. A number of products elaborated by FLSs, including IL-8, IL-16, MCP-1, MIP-1a, and RANK ligand serve as potent chemoattractants, drawing various inflammatory cells from the bloodstream into the joint. These chemoattractants act in concert with other proinflammatory cytokines, such as IL-1β, IL-6, TNF−α, IL-11, IL-15, RANK ligand, and MIF, as well as PGE2, to render the vasculature more permeable, as well as adherent, to inflammatory cells [56,57]. The above characteristics are not observed typically in FLSs cells from patients with osteoarthritis (OA) [48,49,52-55].

PDFG stimulates various cellular functions, including proliferation and differentiation through a cell surface tyrosine kinase receptor [58,59]. It binds and triggers the PDGF-receptor (PDGFR) α and β chains. Both PDGF and PDGFRs are upregulated in RA synovial membrane suggesting a role for PDGF in synovial fibroblast proliferation [12,60-61]. Elevated levels of PDGF have also been detected in synovial fluid of RA patients [62]. In addition to promoting proliferation, PDGF has been shown to increase the expression of proinflammatory cytokines and matrix metalloproteinase-1 in synovial fibroblasts [61,63]. PDGFRs are expressed abundantly on the surface of FLSs in RA, and stimulation with PDGF enhances both the anchorage-dependent and -independent growth of these cells [12,13,52,64-69], implicating PDGF in the activation and transformation of FLSs. Recently, Sandler et al [70], demonstrated that PDGF stimulate the proliferation of synovial fibroblasts in a dose-dependent manner.

PDGF consists of four polypeptide chains that form five disulphide-bonded dimeric PDGF isoforms: AA, AB, BB, CC and DD. PDGFR, in turn, may occur as α or β homodimers or as α/β heterodimers [71]. The binding of the dimeric ligand to the extracellular portions of the two PDGFR chains elicits the dimerization of PDGFRs and the autophosphorylation of their cytoplasmic tyrosine residues. Consequently, multiple *src*-homology 2 (SH2) domain containing signal transduction molecules are recruited to the cell surfaces, including adapter proteins Grb2, Nck and Shc. Beside, accumulating evidence demonstrates the necessity of adapter proteins, including the Gab family, in the PDGF signalling pathway [58,59,72-74]. Recently, Kameda et al [75] showed that both Gab1 [110–120 kDa) and Gab2 [97–100 kDa) exhibited rapid tyrosine phosphorylation that peaked at 1 min after PDGF stimulation and decayed gradually thereafter in FLS from RA patients, suggesting that the adapter proteins Gab1 and Gab2 were thought to be involved in the PDGF-R signalling pathway in RA-FLSs.

Paniagua et al [32], demonstrated that imatinib at concentrations as low as 0.25 μM inhibited PDGF-BB–induced fibroblast proliferation. Imatinib at 0.5 and 5 μM also potently inhibited phosphorylation of PDGFRβ as well as the downstream signalling molecule Akt.

On the other hand, Koyama et al [31] showed that the expression of PDGF-A in the synovial tissue was up-regulated after the induction of arthritis, which was reduced by either the preventive or the therapeutic treatment with 1 mg/kg imatinib on mice with arthritis induced by anticollagen type II antibody (CAIA).

Besides the participation of tyrosine kinases, FLSs could be activated through of other kinases, for example, IL-1, basic fibroblast growth factor (bFGF), fibroblast growth factor 2 (FGF-2) and macrophage migration inhibitory factor (MIF), stimulate the proliferation of FLSs through mitogen-activated protein kinases (MAPK) pathway [for review see ref. 76]. In concordance with that, Pakozdi et al [77], found that MIF induces RA-FLSs matrix metaloproteinases (MMPs)-2 expression in a time-dependent and concentration-dependent manner, and this required the protein kinase C (PKC), c-jun N-terminal kinase (JNK), and Src signalling pathways.

Moreover, it has been demonstrated that inhibition of MAPK pathways, including p38 MAPK, ERK-1/2, and JNK in different cells from RA patients, prevented the expression of RANKL, the production of MMPs and cytokines, and cartilage destruction [76].

MAPKs constitute a superfamily of serine/threonine proteins kinases involved in the regulation of a numbers of intracellular pathways. The mammalian MAP kinase family includes: a) ERK 1 and ERK 2, often referred as p44 and p42 MAP kinases; b) c-Jun-N-terminal or JNK or stress-activated protein kinases (SAPK); c) p38 group: p38α, p38β, p38γ (ERK-6, SAPK3) and p38δ (SAPK-4); d) ERK-5, also called Big MAP kinase (BMK), and e) ERK-3. Unlike the ERK 1/2, which is activated in response to growth and differentiation factors, the JNKs and p38 MAPK are usually activated in response to inflammatory cytokines, such as tumor-necrosis factor (TNF) and interleukin-1 (IL-1) and a diverse array of cellular stresses including UV light, X-rays, hydrogen peroxide (H_2O_2), heat and osmotic shock, and withdrawal of growth factors [for review see ref. 76]

Finally, Morel et al [78], examined the signal transduction mechanisms by which IL-8 induces VCAM-1 expression in RA-FLSs. They demonstrated that IL-8 induced VCAM-1 expression through Src kinase, PI3-kinase/Akt, and ERK1/2 pathways, but they also described a new signalling cascade involving Src/Ras/Raf/ERK/AP-1 in IL-18-stimulated RA-FLSs.

Mast Cells and c-Kit

Increasing evidence suggests that mast cells (MCs), are involved in the pathogenesis of chronic inflammatory diseases and in particular in RA. MCs reside in connective tissues and in synovial tissue of joints. MCs are most prominent beneath the synovial lining layer, in the subintimal areas, and capsule. However, they can be found throughout the synovial tissue. It is very important to remark that MCs can also be found at the sites of cartilage erosions. In RA synovial tissue, MCs comprise up to 5% of the immune cells. They produce an array of proinflammatory mediators, tissue destructive proteases, and cytokines, most prominently TNF-α which is one of the key cytokines in the pathogenesis of RA [for reviews see ref. 79]. MCs can also activate the adaptive immune system by recruiting and presenting antigens to T cells and skewing the balance toward the T-helper 1 (Th1) type of response [80-83]. For the growth and survival of human MCs, stem cell factor (SCF) (also named c-Kit ligand, steel factor, or MC growth factor) is a key growth factor. It is a glycoprotein existing in both soluble and membrane bound forms, after alternative splicing and proteolytic cleavage [84]. In synovial tissue, SCF is expressed by synoviocytes, in particular if they are exposed to TNF-α [85]. MC-derived TNF-α induces the expression of SCF in synovial fibroblasts, which in turn increases the survival of MCs and stimulates their proliferation. In addition to being a growth and survival factor for MCs, SCF is also chemotactic for MCs; it enhances adhesion of MCs to matrix, stimulates MCs to degranulate, and potentiates IgE-induced degranulation [86-88]. The receptor for SCF is c-Kit, which is a tyrosine kinase receptor encoded at human chromosome 4q11–q112 [89]. c-Kit is principally expressed on haematopoietic stem cells, melanocytes, and MCs.

c-Kit expression is downregulated by various pro-inflammatory signals, such as IL-1β, TNF-α, and PMA, in human endothelial cells [90]. GM-CSF, IL-4 and IL-10 downregulate c-Kit expression in the human mast cell line HMC-1 [91-93]. Thus, inhibition of c-Kit in synovial tissue would target relatively specifically MCs. In concordance with that, recently Juurikiyi et al [10] showed that the inhibition of c-Kit by its inhibitor imatinib mesylate leads to significant reduction of TNF-α production in synovial tissue cultures, as well as apoptosis in cultured mast cells, which is probably a key mechanism of the efficient rheumatoid arthritis treatment with anti-TNF-α [94]. They obtained synovial tissue from RA patients and studied the effect of inhibiting c-Kit in synovial tissue explant cultures. Culture of rheumatoid synovial tissue for 24 h in the presence of imatinib clearly increased the number of apoptotic MCs. Whether this inhibition was caused by direct inhibition of TNF-αsynthesis/release or was the result of apoptosis and the inability of dying MCs to synthesize TNF is not clear.

Paniagua et al [32], characterized the effects of imatinib on activation of the cloned murine mast cell line C1.MC/57.1. These cells expand in a growth factor independent fashion, and although growth does not depend on SCF, C1.MC/57.1 mast cells respond to SCF by secreting cytokines. Imatinib at 1 and 5 μM dramatically reduced mast cell production of TNF-α, GM-CSF, and IL-6 to levels similar to those in the unstimulated cell populations. They also demonstrated that imatinib potently inhibited SCF-induced phosphorylation of c-Kit, with a corresponding reduction in phosphorylation of downstream Akt, members of MAPK pathways including ERK, JNK, and p38, and Raf.

Macrophage, c-Fms, and VEGFR

Macrophages appear to play a pivotal role in RA because they are numerous in the inflamed synovial membrane and at the cartilage–pannus junction. They show clear signs of activation, such as overexpression of major histocompatibility complex class II molecules, proinflammatory or regulatory cytokines and growth factors including IL-1, IL-6, IL-10, IL-13, IL-15, IL-18, TNF)-α and GM-CSF, chemokines and chemoattractants such as IL-8, macrophage inflammatory protein (MIP)-1 and monocyte chemoattractant protein (MCP)-1, MMPs and neopterin (for review, see ref. 95].

The degree of macrophage infiltration/activation correlates not only with the joint pain and general inflammatory status of the patient, but also with the radiological progression of permanent joint damage, the disease feature that ultimately determines quality of life [95].

Macrophage expresses the colony-stimulating factor-1 (CSF-1) receptor tyrosine kinase (CSF-1R). CSF-1 is the most pleiotropic macrophage growth factor, stimulating the survival, proliferation and differentiation of mononuclear phagocytes, in addition to the spreading and motility of macrophages [96]. It has been demonstrated that administration of antibodies to CSF-1 reduces the severity of CIA in mice [97].

The CSF-1R is encoded by the c-*fms* proto-oncogene. It contains an extracellular ligand-binding domain, a transmembrane domain, and an intracellular tyrosine domain that is interrupted by a kinase insert domain. CSF-1 is a disulfide-linked homodimer that, by binding CSF-1R, stabilizes CSF-1R dimerization to activate the receptor through autophosphorylation in *trans*, thereby initiating a series of membrane-proximal tyrosine phosphorylation cascades, leading to rapid stimulation of cytoskeletal remodelling, gene transcription and protein translation [96]. On the other hand, GM-CSF stimulates progenitor cells to proliferate and to differentiate to both granulocytes and macrophages, and it also stimulates the proliferation of macrophages.

Recently, Paniagua et al [32], demonstrated that imatinib potently inhibits M-CSF–induced macrophage activation through c-Fms. It inhibited M-CSF–induced phosphorylation of c-Fms, and blocked phosphorylation of protein tyrosine kinases in the MAPK family and other pathways downstream of c-Fms, including Akt (Ser473 and Thr308), ERK, JNK, and p38.

Vascular endothelial growth factors (VEGFs) and their receptors (VEGFRs), including VEGFR-1 (Flt-1), VEGFR-2 (KDR/Flk-1), and VEGFR-3 (Flt-4), form a regulatory system crucial for normal development and pathological angiogenesis [98-101]. VEGFRs are structurally related to the Fms/Kit/PDGFR family and contain an extracellular domain carrying 7 immunoglobulin (Ig)–like repeats and a cytoplasmic tyrosine kinase domain [102,103]. VEGFR-1 and VEGFR-2 are highly expressed on vascular endothelial cells, fibroblasts, but also in monocytes/macrophages and its signaling is involved in the migration of macrophages toward VEGF-A [104-110]. These receptors are thought to mediate different cellular responses (VEGFR1 is primarily involved in vasculogenesis and VEGFR2 in endothelial cell differentiation) possibly by activating different intracellular signalling pathways.

In RA VEGF-A is strongly expressed in synovial macrophages, fibroblasts surrounding microvessels, and vascular smooth muscle cells [111,112]. Recently, Murakami et al [113], have shown that VEGFR-1 tyrosine kinases signals contribute to the symptoms of arthritis, including pathological findings, in a gene-dosage-dependent manner and confirmed the therapeutic effect of VEGFR kinase inhibitors on arthritis, when administered such an inhibitor, KRN951, to mice with *pX*-induced chronic arthritis and collagen-antibody–induced acute arthritis. The administration of KRN951 in the acute model also reduced the progression of the symptoms of arthritis in a dose dependent manner. They suggested that VEGFR tyrosine kinases, particularly VEGFR-1, contribute to the progression of arthritis. Consistent with those finding, De Bandt et al [114] reported that VEGFR-1 is involved in a model of chronic arthritis in which the K.RN/NOD (αvβ6 T-cell receptor [TCR] transgene)/nonobese diabetic) mice were used for the induction of inflammation. Finally, pathological up regulation of the VEGF/VEGFR2 pathway resulting in activated angiogenesis has been demonstrated in the synovium of patients with RA [115].

Grosios et al [116], showed that PTK787/ZK222584 a receptor tyrosine kinase inhibitors with specific activity against the VEGFRs, and that exhibits strong inhibition of VEGF-R2 (KDR) and slightly weaker inhibition of VEGFR1 (Flt-1), Flk-1 (the mouse homologue of KDR) and Flt-4 (the receptor found in the lymphatic system), inhibited knee swelling by 40% in CIA, at the dose of 30 mg/kg, once daily and inhibited both severity scores (by 51%) and global histological scores in mice with collagen-induced arthritis following oral treatment (45 mg/kg), as compared to control animals. PTK787/ZK222584 demonstrated no effects on inflammatory mediators in the VEGF-independent rat carrageenan model and displayed interesting analgesic activity in the Randall Selitto test in the acute setting.

Recently, Koyama et al [31], showed that the expression of VEGF in the synovial tissue was upregulated after the induction of arthritis, which was reduced by either the preventive or the therapeutic treatment with 1 mg/kg imatinib on mice with arthritis induced by anticollagen type II antibody (CAIA).

Osteoclast C-Fms, and C-Src

Destruction of bone and cartilage is one of the most serious complications in RA patients, which can result in progressive morbidity and significant disability. The cells involved in this complication have been extensively studied and although the mechanisms of cartilage destruction in RA were well described; the specific mechanisms responsible for bone erosion in this disease have only recently been studied [76]. Recent studies have shown that osteoclasts are involved in the pathogenesis of bone and joint destruction and can be a potent therapeutic target of this disease.

Osteoclasts are multinucleated, terminally differentiated cells that degrade mineralized matrix during normal and pathological bone turnover [117]. Osteoclastic bone resorption involves the proliferation and homing of the hemopoietic osteoclast progenitors to bone, their differentiation and fusion to form multinucleated cells, and the migration of osteoclasts to and between the resorption sites.

Src family tyrosine kinases are implicated in various intracellular signalling pathways, including mitogenic response to growth factors in fibroblasts, activation of lymphocytes, and osteoclastic bone resorption. Non-receptor type tyrosine kinase c-Src is a member of a family of nine proteins tyrosine kinases that is associated with the cytoplasmic surface of the cellular membrane [118]. Consequently, Src family kinases are implicated in a variety of pathological events in RA, characterized by the proliferation of activated synoviocytes, abnormal immune responses, and elevated production of proinflammatory cytokines.

It has been recently reported that down- or up-regulation of c-Src activity modulates osteoclastic bone resorption not only in vitro but also in vivo [119]. Soriano et al [120] reported that targeted disruption of the c-*src* gene in mice induced osteopetrosis, a disorder characterized by decreased bone resorption, without showing any obvious morphological or functional abnormalities in other tissues or cells.

In concordance with that, a reduction of Src kinase activity has been associated with an inhibition of osteoclast function [121,122].

The kinase activity of Src family kinases is negatively regulated by another cytoplasmic tyrosine kinase, Csk (C-terminal Src kinase), by the phosphorylation of tyrosine residue at the carboxyl-terminus (Tyr 527) [123,124].

Takayanagi et al [122], demonstrated that using an adenovirus-mediated over-expression of Csk, which decrease the Src kinase activity, dramatically decreased the proliferation rate and IL-6 production of human rheumatoid synovial cells in culture; also, the bone-resorbing activity of human osteoclasts formed in coculture of rheumatoid synoviocytes and PBMCs was remarkably suppressed by Csk expression.

Finally, some investigators have suggested that by inhibiting signal transduction through c-fms, the tyrosine kinase inhibitors could inhibit bone resorption, and thereby inhibiting osteoclastogenesis [125]. C-fms is a macrophage colony stimulating factor (M-CSF) receptor and is expressed in osteoclast precursor cells. It has been demonstrated that therapeutic concentrations of imatinib inhibit c-fms kinase activity in cultures stimulated by M-CSF [22].

T Cells and TCR Signalling

Abnormal T-cell function has long been thought to play a key role in the pathogenesis of RA, stimulating both the production of pro-inflammatory cytokines and recruitment of other inflammatory cells, resulting in joint destruction and systemic disease. In the T-cell activation process, CD4+ T-cells are engaged by antigenic peptide fragments in a complex with HLA class II molecules, in addition to co-stimulatory molecules, such as CD80/CD86, expressed on the surface of professional antigen presenting cells. The strongest evidence supporting a role for CD4+ T cells in disease pathogenesis is the association between RA and HLA-DRB1; however, the functional role of this association has to be defined [126]. Evidence also supports a role for T-helper (Th) cells, Th17 cells, and impaired CD4+CD25(hi) regulatory T cell function in the pathogenesis of RA [127,128].

Evidence' supporting the role of T cells on the pathogenesis of RA; have been provided by animal models. For example, in CIA, pathogenic Th1 cells that elicit severe inflammatory synovitis mediate the immune response to a joint antigen. Spontaneous arthritis in K/BxN T-cell-receptor transgenic mice is related to an adaptive immune response against a ubiquitous protein whose end-stage effectors mechanisms are heavily dependent on the innate immune system. In the SKG model of autoimmune inflammatory arthritis, a point mutation in the gene encoding a key signal-transduction molecule in T cells causes defective T cell selection in the thymus, which releases polyclonal autoreactive T cells [128]. Additionally, the new drug studies demonstrating efficacy of agents against T cell, might represent the strongest support for a role of these cells in RA. This is the case with abatacept (a fusion protein--cytotoxic T-lymphocyte-associated antigen 4-IgG1 (CTLA) which is the first selective co-stimulation modulator approved by the US Food and Drug Administration for the management of RA. It is a fusion protein developed to modulate the T-cell co-stimulatory signal that is mediated through the CD28-CD80/86 pathway. The evidence suggests that abatacept is effective in controlling symptoms and improving HRQOL in patients with active RA and an inadequate response to DMARD therapy [129].

The T cell receptor signalling complex (TCR) recognition of self and foreign peptides presented by major histocompatibility complex (MHC) proteins is the crucial event determining whether T cells will become activated. There are four main families of protein tyrosine kinases associated with TCR signalling: Src family kinases (SFK), ZAP 70/Syk, Tec, and Csk [130]. Two SFK members, Lck and Fyn, are involved in the very earliest steps of TCR activation; and Lck is suggested to be more important for TCR signaling [131]. Thus, for all above explained, targeting also SFK activity in T-cell could be useful in the treatment of RA.

Recently, Schade et al [27], demonstrated that dasatinib inhibits TCR-mediated signal transduction, cellular proliferation, cytokine production, and in vivo T cell responses. However, dasatinib-mediated inhibition does not induce apoptosis, as the effect is reversible or may be overcome by signals bypassing the TCR, such as phorbol ester. Signal transduction and proliferative responses via IL-2 remain essentially unperturbed, suggesting dasatinib displays specificity for TCR signalling.

In another study, Leder et al [132], showed that imatinib inhibits antigen-specific IFN-γ secretion of both CD4+ and CD8+ T-effector cells at therapeutically relevant concentrations, while T cells remain responsive. The decrease of IFN-γ production was not due to the loss of T-cell viability.

Others Tyrosine Kinases in RA Synovial Cells

As it was mentioned previously, cytokines identified as major contributors in the hyperplasia of rheumatoid synovial cells include epidermal growth factor (EGF), platelet-derived growth factor (PDGF), and fibroblast growth factor [133].

Although the synovial fluid of RA tissue contains high concentrations of EGF [134], there is very little information about the receptor for EGF in this tissue. In addition to EGF receptor (EGFR), ErbB-2 (also designated c-neu or HER2), ErbB-3, and ErbB-4 (the products of the *erb* B-2, *erb* B-3, and *erb* B-4 genes, respectively) can also function as receptors for the EGF family of growth factors [135-138]. Satoh et al [139], showed that expression of ErbB-2, but not other ErbB-related products were detected in synovium with RA as compared with that with OA and ligament injury. Genistein, a tyrosine kinase inhibitor, and herceptin, a specific monoclonal antibody against ErbB-2 inhibited growth of primary synovial cells with RA. Herceptin showed a small effect on growth of primary synovial cells with OA. EGF stimulated the phosphorylation of ErbB-2 in primary synovial cells with RA. This EGF-stimulated phosphorylation was completely abrogated by genistein and herceptin.

Janus kinases (JAK) are other groups of tyrosine kinases involve in RA. JAK family of tyrosine kinases includes JAK1, JAK2, JAK3 and TYK2, and is required for signaling through Type I and Type II cytokine receptors.

Several cytokines implicated in RA pathogenesis, including IL-6, IL-15, granulocyte–macrophage colony-stimulating factor, and interferons (IFNs) activate the JAK/STAT pathway, and STAT-3 is constitutively activated in RA [140]. The JAK/STAT signal transduction pathway is utilized by many cytokines and growth factors that regulate gene expression and cellular activation, proliferation, and differentiation [141]. Typically, stimulation by a particular cytokine results in the activation of a distinct pair of 2 of the 4 known JAKs. JAK kinases are required for tyrosine phosphorylation and activation of latent cytoplasmic transcription factors termed signal transducers and activators of transcription (STATs).

In concordance with that, Migita et al [142], showed that pre-treatment of FLSs, from RA patients, with CP-690550 (a JAK inhibitor) and stimulated with recombinant oncostatin M (OSM), completely abrogated the OSM-induced production of IL-6, as well as OSM-induced JAK/STAT, and activation of MAPKs. These findings suggest that IL-6-type cytokines contribute to rheumatoid synovitis through activation of the JAK/STAT pathway in rheumatoid synoviocytes. Additionally, it had been showed that CP-690550 potently inhibited signaling through JAK1 and JAK3 with 5-100 fold selectivity over JAK2 in cellular assays, despite inhibiting all four JAK isoforms with nM potency in *in vitro* enzyme assays. Plasma cytokines (IL-6 and IL-17), PBNC, and bone marrow myeloid progenitor cells returned to pre-disease levels [43]. Finally, the INCB028050 a selective orally bioavailable JAK1/JAK2 inhibitor with nanomolar potency against JAK1 (5.9 nM) and JAK2 (5.7 nM) inhibits intracellular signalling of multiple proinflammatory cytokines including IL-6 and IL-23 at concentrations, 50 nM Inhibition of JAK signaling prevents the production of chemokines and cytokines associated with uncontrolled RA [144].

RA is an inflammatory disease as well as an angiogenic disease, and persistent angiogenesis is critical both to maintaining the chronic architectural changes in the RA synovium via delivery of nutrients and inflammatory cells [145]. Tie2 is an endothelium-specific receptor tyrosine kinase required for embryonic vascular development. Disruption of Tie2 function in transgenic mice results in embryonic death due to vascular defects [146,147]. Angiopoietin 1 (Ang1) is an agonist ligand for Tie2. It stimulates Tie2 phosphorylation and activates Tie2 signalling [148]. Tie2 protein and Ang1 messenger RNA have been detected in different diseases, including RA synovium [149,150]. DeBusk et al [145], implicated Tie2 in pathologic angiogenesis. They observed that Tie2 and Ang1 were elevated in human RA synovium. Using a novel collagen-induced arthritis synovial window model, they demonstrated that Tie2 signalling regulated arthritis angiogenesis in vivo. In addition, they observed that TNF-α can regulate Tie2 activation in multiple ways that may involve interactions between endothelial cells and synoviocytes. TNF-α up-regulates Tie2 in endothelial cells through nuclear factor κB, and it up-regulates Ang1 in synoviocytes. In concordance with that finding and using a novel synovium vascular window model, they demonstrated that blockade of Tie2 activation (AdExTek) inhibited angiogenesis in vivo.

Recently, it has been involved another TK in the pathogenesis of RA [151]. Spleen tyrosine kinase (Syk) is an intracellular protein tyrosine kinase that serves as a key mediator of Fc receptor and B cell receptor signalling in inflammatory cells, including mast cells, macrophages, dendritic cells (DC), NK cells, and neutrophils [151]. This protein kinase associates with a variety of receptors on the surface of different cells of the immune system [152]. The binding of a Syk-associated receptor (such as the Fc receptor) to its ligand leads to the phosphorylation of a specialized protein motif called immunoreceptor tyrosine activating motif (ITAM) by the Src-family kinases [153].

It has been showed that Syk signalling is implicated in TNFα-induced inflammatory cytokine and MMP production by FLS isolated from joint tissues of RA patients [154]. Additionally, the treatment of cultured FLS with R406 markedly suppressed TNFα-induced IL-6 and MMP-3 production in RA FLS. R406 has been examined in several cell systems, and its activity is consistent with Syk inhibition in isolated mast cells, macrophages, DC, and synoviocytes [151]. Therefore, inhibition of Syk can prevent bone erosions by dampening the inflammatory response (via a decrease in the production of cytokines such as IL-1 and TNF-α, and inhibition effector T cell function) or by directly inhibiting bone resorption by osteoclasts [153].

In concordance with that, Pine et al [151] showed that treatment of arthritic rats with either R406 or R788 (prodrug), resulted in a significantly reduced serum cartilage oligomeric matrix protein (COMP), and synovial cytokines expression. COMP, a cartilage degradation product in the serum, is associated with loss of bone integrity in RA patients [155,156]. Expression of Syk in synovial tissue correlated with the joint level of inflammatory cell infiltrates and was virtually undetectable in treated rats.

Clinical and Animal Studies with TKIs in RA

Clinical Studies

There are few clinical studies of TKIs in RA patients. Initilly, Eklund et al [28], reported a patient with gastrointestinal stromal tumour and RA who was treated with imatinib mesylate. After starting the imatinib treatment, she reported that her concomitant RA symptoms clearly improved. After that, they treated three patients with severe RA with imatinib (400 mg/day) for 12 weeks. All previous medications used by these patients, including methotrexate and the biological anti-TNF treatment, had failed.

The primary outcome measures were: numbers of tender and swollen joints, patient-assessed disease activity and pain as assessed by a visual analogue scale, a health assessment questionnaire (HAQ) score, serum C-reactive protein (CRP), and blood erythrocyte sedimentation rate (ESR); all of them improved during imatinib therapy. The swollen joint count decreased in all patients and the tender joint count in two of the three patients. HAQ score improved and the patients reported less pain and disease activity. CPR and ESR improved in two patients and one interrupted therapy due to a rash. One of these patients had no good treatment options left, and as he responded to treatment, he continued with imatinib for nearly 3 years. During the follow-up, his disease activity was constantly low. After 4 months of treatment, he fulfilled the stringent 70 improvement criteria of the American College of Rheumatology at most visits. In another case when the patient had concomitant RA and chronic myelogenous leukaemia (CML) the treatment with imatinib improved both of diseases [29]. He was an 80 year-old man patient that was treated for CML with 300 mg of imatinib mesylate for 8 weeks. Laboratory tests showed that the C-reactive protein level, percentage of cells exhibiting the Philadelphia chromosome (Ph1), WBC count, and Lansbury index for RA all dropped respectively from 7.5 mg/dl to 1.0 mg/dl, 74.9% to 1%, 25, 100/μl to 9900/μl, and 51% to 14%.

In a 12-week, ascending-dose, randomized, placebo-controlled trial Wienblatt et al [157], found in 189 patients with active RA despite methotrexate therapy, a significant reduction in arthritis activity and in serum levels of interleukin-6 and MMPs 3, in the two groups that received the highest doses of R788 (Syk Kinase Inhibitors) (100 mg twice daily and 150 mg twice daily), as compared with the groups that received placebo or the 50-mg dose of R788 twice daily.

After that, a larger and longer phase 2 trial was designed to evaluate R788 in 457 patients with active RA no responding to methotrexate therapy (R788 in 152 patients at a dose of 150 mg once daily and 152 at a dose of 100 mg twice daily and 153 received placebo, 77 patients once daily and 76 twice daily)[158]. The primary outcome was the rate at 6 months of the ACR 20 response, which is defined as at least a 20% reduction from baseline in the number of both tender and swollen joints and a 20% or greater improvement in at least three of five measures, and the secondary outcomes were the rates of ACR 50 and ACR 70 responses, improvements in individual components of the ACR score, including the HAQ; disease activity (DAS28), remission of rheumatoid arthritis (DAS28 score below 2.6); the score on the Functional Assessment of Chronic Illness Therapy (FACIT)– fatigue scale; and scores on the Medical Outcomes Study 36-Item Short- Form Health Survey (SF-36).

They found that R788, at a dose of 100 mg twice daily and at a dose of 150 mg once daily, was significantly superior to placebo at month 6 (ACR 20 response rates of 67% and 57%, respectively, vs. 35%; $P<0.001$ for the comparison of both doses with placebo). It was also significantly superior with respect to ACR 50, which indicates at least a 50% improvement (43% and 32% vs. 19%; $P<0.001$ for the comparison of the 100-mg dose with placebo, $P = 0.007$ for the comparison of the 150-mg dose with placebo) and ACR 70 (28% and 14% vs. 10%; $P<0.001$ for the comparison of the 100-mg dose with placebo, $P = 0.34$ for the comparison of the 150-mg dose with placebo).

The DAS28 was also significantly improved by month 1 in the group receiving 100 mg of R788 twice daily, and the improvement was maintained throughout the study. At month 6, the rates of DAS28 below 2.6, were 21% in the group receiving R788 at a dose of 150 mg once daily and 31% in the group receiving 100 mg of R788 twice daily, as compared with 7% in the placebo group ($P = 0.003$ for the comparison of the 150-mg dose with placebo, $P<0.001$ for the comparison of the 100-mg dose with placebo).

Adverse effects included diarrhea (in 19% of subjects taking the 100-mg dose of R788 vs. 3% of those taking placebo), upper respiratory infections (14% vs. 7%), and neutropenia (6% vs. 1%). R788 was associated with an increase in systolic blood pressure of approximately 3 mm Hg between baseline and month 1, as compared with a decrease of 2 mm Hg with placebo; 23% of the patients taking R788 vs. 7% of the patients receiving placebo required the initiation of or a change in antihypertensive therapy.

More recently in a three-month randomized, placebo-controlled, phase II study in 129 patients with active RA that did not respond to biologic agents, Genovese et al [159], found no differences in the primary end point between the R788 and placebo groups. However, there were baseline differences in steroid use, prior biologic use, and synovitis score on MRI between the R788 group and the placebo group that may have affected the outcomes.

The ACR20 response in the R788 100 mg twice daily group was 38%, versus 37% in the placebo group, at month 3. No significant differences were achieved in the ACR20, ACR50, or ACR70 response levels at 3 months. There were differences between the groups from baseline to month 3 in the secondary end points C-reactive protein (CRP) level and synovitis score on MRI. These controversial results with the Syk Kinase Inhibitor (fostamatinib) will need to be evaluated in further clinical studies.

Recently, have been evaluated the effect of JAK inhibitors in RA patients. Kremer et al [160], determined the efficacy, safety, and tolerability of 3 different dosages of tofacitinib (CP-690550), a potent, orally active JAK inhibitor, in patients with active RA in whom methotrexate, etanercept, infliximab, or adalimumab caused an inadequate or toxic response. 264 patients were randomized equally to receive placebo, 5 mg of CP-690550, 15 mg of CP-690550, or 30 mg of CP-690550 twice daily for 6 weeks, and were followed up for an additional 6 weeks after treatment. The primary efficacy end point was the ACR20 response rate at 6 weeks. They found that the ACR20 response rates were 70.5%, 81.2%, and 76.8% in the 5 mg, 15 mg, and 30 mg twice daily groups, respectively, compared with 29.2% in the placebo group ($P < 0.001$). Improvements in disease activity in CP-690550-treated patients compared with placebo were seen in all treatment groups as early as week 1. ACR50 and ACR70 response rates significantly improved in all treatment groups by week 4. The most common adverse events reported were headache and nausea. The infection rate in both the 15 mg twice daily group and the 30 mg twice daily group was 30.4% (versus 26.2% in the placebo group). No opportunistic infections or deaths occurred. Increases in mean low-density lipoprotein cholesterol and high-density lipoprotein cholesterol levels, and increases in mean serum creatinine level (0.04-0.06 mg/dl) were seen in all CP-690550 treatment arms. Coombs et al [161], evaluated the efficacy of CP-690550 in improving pain, function and health status in patients with moderate to severe active RA and an inadequate response to methotrexate or a tumour necrosis factor alpha inhibitor. They showed patients treated with CP-690550 (5, 15 and 30 mg twice-daily groups) experienced a 50% improvement in pain compared with placebo (44%, 66%, 78% and 14%, respectively), clinically meaningful reductions in HAQ-DI (> or =0.3 units) (57%, 75%, 76% and 36%, respectively) and clinically meaningful improvements in SF-36 domains and physical and mental components. Cohen et al [162], showed that the co-administration of CP-690550 (30 mg b.i.d.) and MTX (15-25 mg per week) in twelve patients diagnosed with RA, was safe and well tolerated. There was no clinically significant effect on the pharmacokinetics profile of either drug.

More recently, Kremer et al [163], compared the efficacy, safety adnd tolerability of six doses of oral tofacitinib (CP-690550) with placebo for the treatment of active RA in patients on a stable MTX with inadequate response to MTX monoptherapy. In a 24-week, doible-blind, Phase 2B study, patients with active RA (N=507) and stable background of MTX, were randomized to tofacitinib 1,3, 5, 10, 15 mg twice daily, 20 mg once or placebo. At Week 12, ACR20 response rates were significantly ($p \leq 0.05$) greater than placebo (33.3%) for all tofacitinib doses ≤3 mg: 52.9% (3 mg), 50.7% (5 mg), 58.1% (10 mg), 56.0% (15 mg), and 53.8% (20 mg), respectively. Improvements were sustained at Week 24 for ACR20, ACR50, ACR70, Health Assessment Questionnaire-Disability Index, DAS 28-3, and DAS28-3 (CRP)<2.6. The most common treatment-emergent adverse events (AEs) occurring in >10% of patients in any tofacitinib group were diarrhea, upper respiratory tract infection, urinary tract infection, arthralgia, and headache; 21 patients (4.1%) experienced serious AEs. Sporadic increases in transaminases, increases in cholesterol and serum creatinine, and decreases in neutrophils and hemoglobin were observed.

Animal Studies

In an early study, Takayanagi et al [122] showed that by repressing Src kinase activity through adenovirus-mediated overexpression of Csk, the direct injection of Csk virus into rat ankles with adjuvant arthritis not only ameliorated the inflammatory reactions but also suppressed bone destruction.

Paniagua et al [32] showed that mice treated with either 33 or 100 mg/kg imatinib displayed significant reductions in the severity of CIA based on reduced paw swelling, erythema, and joint rigidity as assessed by the mean visual arthritis score and reduced mean paw thickness ($P < 0.01$ by Mann-Whitney U test for both the 33 and 100 mg/kg groups after day 38 following primary immunization). Imatinib also reduced the incidence of CIA. The therapeutic effects of imatinib demonstrated a trend toward dose dependence. On the other hand, in mice with established clinical arthritis (average visual score of 4) were randomized and treated with 33 or 100 mg/kg imatinib or PBS. Both the 33 and 100 mg/kg dose levels of imatinib inhibited the progression of established arthritis as assessed by both the visual scoring system and mean paw thickness ($P < 0.05$ or $P < 0.01$ after 10 days following the initiation of treatment). They also performed histopathologic evaluation, and demonstrated that imatinib resulted in statistically significant reductions in synovitis, pannus, and erosion scores in both the CIA prevention and established CIA treatment ($P < 0.05$ for synovitis, pannus, and erosions scores) experiments. Thus, imatinib was effective at both preventing and treating established CIA based on clinical and histopathologic analyses.

Recently, Koyama et al [31], showed that mice treated with intraperitoneal administration of imatinib (1 or 10 mg/kg) prior to the development of CAIA displayed significant reductions in the severity of CAIA as assessed by arthritis score, histology, and synovial PDGF and vascular endothelial growth factor expression. In addition, treatment of the mice that had developed CAIA with intraperitoneal administration of imatinib (1 or 10 mg/kg) inhibited the progression of arthritis as assessed by those parameters. These results suggest that imatinib prevents and treats CAIA.

More recently, Akashi et al [164], compared the effects of TKIs imatinib and nilotinib in models of autoimmune arthritis. They investigate whether imatinib or nilotinib was effective against arthritis in the glucose-6-phosphate isomerase (GPI)-induced arthritis mouse model. Imatinib or nilotinib was administered orally to the arthritic mice at different time points. Efficacy was evaluated by visual scoring and by determining the production of anti-GPI antibody. Splenocytes from the arthritic mice were cultured with GPI in the presence of imatinib or nilotinib in vitro, and cytokine levels in the culture supernatants were analyzed. To investigate the effects of imatinib and nilotinib on T-cell proliferation, lymph node cells from the arthritic mice were cultured with GPI in the presence of imatinib or nilotinib in vitro. IL-17 mRNA expressions in the arthritic ankle joints from the onset of arthritis was analyzed by real-time polymerase chain reaction (PCR). The administration of imatinib from day 0 showed suppression of arthritis, the administration of nilotinib from day 0 resulted in pronounced suppression of arthritis and that from day 7 showed significant inhibition of the progression of arthritis. A reduction in anti-GPI antibodies was correlated with the therapeutic efficacy of imatinib, but not with that of nilotinib. Imatinib dose-dependently inhibited TNF-α, IL-6, IFN-γ, and IL-17 production by splenocytes in vitro, while nilotinib inhibited only IL-17 and IFN-γ production in a dose-dependent fashion. Imatinib at 3 μM exerted a mild antiproliferative effect on CD4+ T cells ($P < 0.05$), whereas imatinib at 10 μM and nilotinib at

3 and 10 μM demonstrated a marked antiproliferative effect ($P < 0.01$). The IL17 gene expression level on day 7 tended to be higher than that on day 14. These findings suggest that imatinib and nilotinib could prevent autoimmune arthritis, essentially via distinct mechanisms, in that imatinib inhibits both inflammatory and T-cell-derived cytokine production, whereas nilotinib suppresses T-cell-derived cytokine production. Imatinib and nilotinib could have therapeutic potential for RA and other inflammatory diseases.

Other two TKIs (Syk inhibitors), R406 and R788 have been examined in rat's models of arthritis [151]. First, the investigators showed that prophylactic treatment of rats with R406 or R788 (prodrug) treatment reduced the cutaneous reverse passive Arthus (RPA) reaction and inflammatory edema in a dose-dependent manner. Then, they showed that R406 or R788 treatment of arthritic rats reduce the progression and severity of clinical arthritis that was evident within 7 days after initiation of therapy, and the clinical condition of rats receiving drug therapy at efficacious doses remained improved throughout the study. It was also associated with a bone preservation or retarded destruction. Finally, they found a reduced histologic synovitis and destruction in R788/R406-treated rats.

Recently, other studies have evaluated the inhibitor effects of CP-690550, CEP-33779 and INCB028050, all potent and selective JAK inhibitors. Fridman et al [144], assessed the effects of INCB028050 on rat adjuvant-induced model of arthritis (rAIA). They used oral INCB028050 at daily doses of 1, 3, or 10 mg/kg based on its pharmacokinetic profile in this species. Disease severity was assessed periodically, scoring clinical signs of disease. Relative to vehicle-treated animals, increasing doses of INCB028050 inhibited disease scores by 24%, 57%, and 81%, respectively. Moreover, while the observed changes in clinical scores at study termination were impressive, suppression of clinical disease by INCB028050 was observed as early as 2 day after initiation of treatment. INCB028050 treatment, compared with vehicle, inhibited the increase in hind paw volumes during the 2 week of treatment by 50% at a dose of 1 mg/kg and.95% at doses of 3 or 10 mg/kg. Additionally, joint inflammation was inhibited by 27%, 64%, and 82% at doses of 1, 3, and 10mg/ kg, respectively, compared with vehicle controls. This was consistent with respective decrements in ankle width of 19%, 62%, and 77% compared with vehicle controls (0%) and historical naive animals. Finally, they evaluated the disease-modifying potential of INCB028050 by examining histologic evidence of bone resorption. Treatment of diseased animals with INCB028050 reduced bone resorption by 15% 61%, and 67% with increasing dose level (1, 3, and 10 mg/kg) and resulted in radiographic improvements with restoration of the normal architecture and appearance to the ankle and tarsals. Milici et al [165], evaluated CP-690550 in murine collagen-induced (CIA) and rat adjuvant-induced (AA) models of RA. Animals received the JAK3 inhibitor via osmotic mini-pump infusion at doses ranging from 1.5-15 mg/kg/day following disease induction. Treatment at both the intermediate (5 mg/kg/day) and high (15 mg/kg/ day) doses of CP-690550 produced a highly significant, near total suppression of clinical scores throughout the entire study. Based upon the clinical scores, the ED50 of CP-690550 was ~1.5 mg/kg/day with > 90% disease reduction observed at the 15 mg/kg/day dose. The compound also reduced inflammatory cell influx and joint damage as measured histologically. Animals receiving a CP-690550 dose of 15 mg/k/d showed no histological evidence of disease. Meyer et al [143], evaluated the effect of CP-690550 rat adjuvant-induced arthritis. Rats were dosed orally with either vehicle (0.5% methylcellulose/0.025% Tween-20) alone or CP-690550 suspended in vehicle. Dosing was initiated on day 14 and continued twice daily through day 21 post-immunization. The volume of both the right and left hind paw of each rat was measured by plethysmometer and the combined volume of both paws was calculated. CP-690550 treatment resulted in a dose-dependent inhibition of arthritis as indicated by decreased paw volume compared to vehicle treated control rats. Notably, CP-690550 at a 10 mg/kg dose reduced paw volume to within the normal range by day 21. Finally, Stump et al [166], found that CEP-33779, a JAK2 inhibitor, administrated orally twice daily at 10 mg/kg, 30 mg/kg, 55 mg/kg or 100 mg/kg over a period of 4-8 weeks reduced mean paw edema and clinical scores in both CIA and CAIA models of arthritis. Dose-dependent responses were evident by the reduction in mean paw thickness or size over time with treatment (1.3-fold drop at day 13 for 55 mg/kg dose compared to vehicle). This correlated well with a decrease in total clinical score peaking at the highest dose used, 55 mg/kg, (2.4-fold reduction at day 13 for 55 mg/kg dose compared to vehicle). Both models demonstrated histological evidence of disease amelioration upon treatment (e.g., reduced matrix erosions, subchondral osteolysis, pannus formation and synovial inflammation) and reduced paw phosphorylated STAT3 levels.

Conclusions

There are evidences that different tyrosine kinases such of mast cell c-Kit, FLS PDGFR, macrophage c-Fms, Src family kinases, and other kinases have important participation in the physiopathology of RA. Therefore, the inhibition of those kinases may suppose a therapeutic effect. Consistent with these *in vitro* evidences, there are recent cases reports describing severe RA patients treated with TKIs, who showed a marked improvement in the joint pain and disease activity. Beside, animal studies have showed that the TKIs treat, prevent, and inhibit osteoclastogenesis and joint destruction. With all these evidences we can hypothesize that TKIs will be the next step in the treatment of patients with RA.

References

[1] Firestein G.S. *Evolving concepts of rheumatoid arthritis*. Nature 2003;423:356–361.

[2] Choy EHS, Panayi GS. *Cytokine pathways and joint inflammation in rheumatoid arthritis*. N Eng J Med 2001;344:907–16.

[3] Burmester GR, Stuhlmuller B, Keyszer G, Kinne RW. *Mononuclear phagocytes and rheumatoid synovitis. Mastermind or workhorse in arthritis?* Arthritis Rheum 1997;40:5–18.

[4] Kinne RW, Brauer R, Stuhlmuller B, Palombo-Kinne E, Burmester GR. *Macrophages in rheumatoid arthritis*. Arthritis Res 2000;2:189–202.

[5] Woolley DE. *The mast cell in inflammatory arthritis*. N Engl J Med 2003;348:1709–1711.

[6] Woolley DE, Tetlow LC. *Mast cell activation and its relation to proinflammatory cytokine production in the rheumatoid lesion*. Arthritis Res 2000;2:65–74.

[7] Benoist C, Mathis D. *Mast cells in autoimmune disease*. Nature 2002;420:875–878.

[8] Gordon JR, Galli SJ. *Mast cells as a source of both preformed and immunologically inducible TNF-alpha/cachectin*. Nature 1990;346:274–276.

[9] Bromley M, Fisher WD, Woolley DE. *Mast cells at sites of cartilage erosion in the rheumatoid joint.* Ann Rheum Dis 1984;43:76–79.

[10] Juurikivi A. Sandler C, Lindstedt KA, Kovanen PT, Juutilainen T, Leskinen MJ, Mäki T, Eklund KK. *Inhibition of c-kit tyrosine kinase by imatinib mesylate induces apoptosis in mast cells in rheumatoid synovia: a potential approach to the treatment of arthritis.* Ann Rheum Dis 2005;64:1126–1131.

[11] Cheon H, Sun YK, Yu SJ, Lee YH, Ji JD, Song GG, Lee JH, Kim MK, Sohn J. *Platelet-derived growth factor-AA increases IL-1beta and IL-8 expression and activates NF-kappaB in rheumatoid fibroblast-like synoviocytes.* Scand J Immunol 2004;60:455–462.

[12] Watanabe N, Ando K, Yoshida S, Inuzuka S, Kobayashi M, Matsui N, Okamoto T. *Gene expression profile analysis of rheumatoid synovial fibroblast cultures revealing the overexpression of genes responsible for tumor-like growth of rheumatoid synovium.* Biochem Biophys Res Commun 2002;294:1121–1129.

[13] Remmers EF, Sano H, Lafyatis R, Case JP, Kumkumian GK, Hla T, Maciag T, Wilder RL. *Production of platelet derived growth factor B chain (PDGF-B/c-sis) mRNA and immunoreactive PDGF B-like polypeptide by rheumatoid synovium: coexpression with heparin binding acidic fibroblast growth factor-1.* J Rheumatol 1991;18:7–13.

[14] Goldring SR, Gravallese EM. *Pathogenesis of bone erosions in rheumatoid arthritis.* Curr Opin Rheumatol 2000;12:195–199.

[15] Yasuda H, Shima N, Nakagawa N, Yamaguchi K, Kinosaki M, Mochizuki S, Tomoyasu A, Yano K, Goto M, Murakami A, Tsuda E, Morinaga T, Higashio K, Udagawa N, Takahashi N, Suda T. *Osteoclast differentiation factor is a ligand for osteoprotegerin /osteoclastogenesis-inhibitory factor and is identical to TRANCE/RANKL.* Proc Natl Acad Sci U S A 1998;95:3597–3602

[16] Suda T, Takahashi N, Udagawa N, Jimi E, Gillespie MT, Martin TJ. *Modulation of osteoclast differentiation and function by the new members of the tumor necrosis factor receptor and ligand families.* Endocr Rev 1999;20:345–357.

[17] Udagawa N, Takahashi N, Jimi E, Matsuzaki K, Tsurukai T, Itoh K, Nakagawa N, Yasuda H, Goto M, Tsuda E, Higashio K, Gillespie MT, Martin TJ, Suda T. *Osteoblasts/stromal cells stimulate osteoclast activation through expression of osteoclast differentiation factor/RANKL but not macrophage colony-stimulating factor: receptor activator of NF-kappa B ligand.* Bone (NY) 1999;25:517–523

[18] Wood DD, Ihrie EJ, Hamerman D. *Release of interleukin-1 from human synovial tissue in vitro.* Arthritis Rheum 1985;28:853–862

[19] Guerne PA, Zuraw BL, Vaughan JH, Carson DA, Lotz M. *Synovium as a source of interleukin 6 in vitro. Contribution to local and systemic manifestations of arthritis.* J Clin Invest 1989;83:585–592

[20] Ritchlin C. Fibroblast biology. *Effector signals released by the synovial fibroblast in arthritis.* Arthritis Res 2000;2:356–360

[21] Takayanagi H, Iizuka H, Juji T, Nakagawa T, Yamamoto A, Miyazaki T, Koshihara Y, Oda H, Nakamura K, Tanaka S. *Involvement of receptor activator of nuclear factor kappaB ligand/osteoclast differentiation factor in osteoclastogenesis from synoviocytes in rheumatoid arthritis.* Arthritis Rheum 2000;43:259–269.

[22] Dewar AL, Cambareri AC, Zannettino AC, Miller BL, Doherty KV, Hughes TP, Lyons AB. *Macrophage colony stimulating factor receptor, c-fms, is a novel target of imatinib*. Blood 2005;105:3127–3132

[23] van de Wijngaert FP, Tas MC, Burger EH. *Conditioned medium of fetal mouse long bone rudiments stimulates the formation of osteoclast precursor-like cells from mouse bone marrow*. Bone (NY) 1989;10:61–68

[24] Felix R, Cecchini MG, Hofstetter W, Elford PR, Stutzer A. *Impairment of macrophage colony-stimulating factor production and lack of resident bone marrow macrophages in the osteopetrotic op/op mouse. J Bone Miner Res 1990;5:781–789*

[25] Kodama H, Yamasaki A, Nose M, Niida S, Ohgame Y, Abe M, Kumegawa M, Suda T. *Congenital osteoclast deficiency in osteopetrotic (op/op) mice is cured by injections of macrophage colony-stimulating factor*. J Exp Med 1991;173:269–272

[26] Tanaka S, Takahashi N, Udagawa N, Tamura T, Akatsu T, Stanley ER, Kurokawa T, Suda T. *Macrophage colony-stimulating factor is indispensable for both proliferation and differentiation of osteoclast progenitors*. J Clin Invest 1993;91:257–263

[27] Schade AE, Schieven GL, Townsend R, Jankowska AM, Susulic V, Zhang R, Szpurka H, Maciejewski JP. *Dasatinib, a small-molecule protein tyrosine kinase inhibitor, inhibits T-cell activation and proliferation*. Blood. 2008;111(3):1366-77.

[28] Eklund KK, Joensuu H. *Treatment of rheumatoid arthritis with imatinib mesylate: clinical improvement in three refractory cases*. Ann Med 2003;35:1-6.

[29] Miyachi K, Ihara A, Hankins RW, Murai R, Maehiro S, Miyashita H. *Efficacy of imatinib mesylate (STI571) treatment for a patient with rheumatoid arthritis developing chronic myelogenous leukemia*. Clin Rheumatol. 2003;22(4-5):329-32.

[30] Ando W, Hashimoto J, Nampei A, Tsuboi H, Tateishi K, Ono T, Nakamura N, Ochi T, Yoshikawa H. *Imatinib mesylate inhibits osteoclastogenesis and joint destruction in rats with collagen-induced arthritis (CIA)*. J Bone Miner Metab. 2006;24(4):274-82.

[31] Koyama K, Hatsushika K, Ando T, Sakuma M, Wako M, Kato R, Haro H, Sugiyama H, Hamada Y, Ogawa H, Nakao A. *Imatinib mesylate both prevents and treats the arthritis induced by type II collagen antibody in mice*. Mod Rheumatol. 2007;17(4):306-10.

[32] Paniagua RT, Sharpe O, Ho PP, Chan SM, Chang A, Higgins JP, Tomooka BH, Thomas FM, Song JJ, Goodman SB, Lee DM, Genovese MC, Utz PJ, Steinman L, Robinson WH. *Selective tyrosine kinase inhibition by imatinib mesylate for the treatment of autoimmune arthritis*. J Clin Invest. 2006;116(10):2633-42.

[33] van der Geer P, Hunter T, Lindberg RA. *Receptor protein-tyrosine kinases and their signal transduction pathways*. Annu Rev Cell Biol. 1994;10:251-337.

[34] Liu X, Pawson T. *Biochemistry of the Src protein-tyrosine kinase: regulation by SH2 and SH3 domains*. Recent Prog Horm Res. 1994;49:149-60.

[35] Levitzki A, Gazit A. *Tyrosine kinase inhibition: an approach to drug development*. Science 1995;267(5205):1782-8.

[36] Marshall CJ. *Specificity of receptor tyrosine kinase signaling: transient versus sustained extracellular signal-regulated kinase activation*. Cell 1995;80(2):179-85

[37] Schlessinger J. *New roles for Src kinases in control of cell survival and angiogenesis*. Cell. 2000;100(3):293-6. Review.

[38] Traxler P. *Tyrosine kinases as target in cancer therapy-successes and failures*. Expert Opin Ther Targets 2003;7:215-234.

[39] Tibes R, Trent J, Kurzrock R. *Tyrosine kinase inhibitors and the dawn of molecular cancer therapeutics.* Annu Rev Pharmacol Toxicol. 2005;45:357-84.

[40] Ullrich A, Schlessinger J. *Signal transduction by receptors with tyrosine kinase activity.* Cell. 1990;61(2):203-12

[41] Heldin CH. *Dimerization of cell surface receptors in signal transduction.* Cell. 1995;80(2):213-23.

[42] Cantley LC, Auger KR, Carpenter C, Duckworth B, Graziani A, Kapeller R, Soltoff S. *Oncogenes and signal transduction.* Cell. 1991;64(2):281-302.

[43] Schindler C, Darnell JE Jr. *Transcriptional responses to polypeptide ligands: the JAK-STAT pathway.* Annu Rev Biochem. 1995;64:621-51.

[44] Cobb MH, Goldsmith EJ. *How MAP kinases are regulated.* J Biol Chem. 1995;270(25):14843-6.

[45] Weiss A, Schlessinger J. *Switching signals on or off by receptor dimerization.* Cell. 1998;94(3):277-80.

[46] Lemmon MA, Schlessinger J. *Transmembrane signaling by receptor oligomerization.* Methods Mol Biol. 1998;84:49-71.

[47] Briscoe J, Kohlhuber F, Müller M. *JAKs and STATs branch out.* Trends Cell Biol. 1996;6(9):336-40.

[48] Müller-Ladner U, Gay RE, Gay S. *Activation of synoviocytes.* Curr Opin Rheumatol 2000;12:186–94.

[49] Yamanishi Y, Firestein GS. *Pathogenesis of rheumatoid arthritis: the role of synoviocytes.* Rheum Dis Clin North Am 2001;27:355–71.

[50] Reme T, Travaglio A, Gueydon E, Adla L, Jorgensen C, Sany J. *Mutations of the p53 tumour suppressor gene in erosive rheumatoid synovial tissue.* Clin Exp Immunol. 1998;111:353-8.

[51] Jang J, Lim DS, Choi YE, Jeong Y, Yoo SA, Kim WU, Bae YS. *MLN51 and GM-CSF involvement in the proliferation of fibroblast- like synoviocytes in the pathogenesis of rheumatoid arthritis.* Arthritis Res Ther 2006, 8:R170.

[52] Lafyatis R, Remmers EF, Roberts AB, Yocum DE, Sporn MB, Wilder RL. *Anchorage-independent growth of synoviocytes from arthritic and normal joints. Stimulation by exogenous plateletderived growth factor and inhibition by transforming growth factor-beta and retinoids.* J Clin Invest 1989;83:1267–76.

[53] Müller-Ladner U, Kriegsmann J, Gay RE, Gay S. *Oncogenes in rheumatoid arthritis.* Rheum Dis Clin North Am 1995;21:675–90.

[54] Imamura F, Aono H, Hasunuma T, Sumida T, Tateishi H, Maruo S, Nishioka K. *Monoclonal expansion of synoviocytes in rheumatoid arthritis.* Arthritis Rheum 1998;41:1979–86.

[55] Müller-Ladner U, Kriegsmann J, Franklin BN, Matsumoto S, Geiler T, Gay RE, Gay S. *Synovial fibroblasts of patients with rheumatoid arthritis attach to and invade normal human cartilage when engrafted into SCID mice.* Am J Pathol 1996;149:1607–15.

[56] Mor A, Abramson SB, Pillinger MH. *The fibroblast-like synovial cell in rheumatoid arthritis: a key player in inflammation and joint destruction.* Clin Immunol. 2005;115:118-28.

[57] Abeles AM, Pillinger MH. *The role of the synovial fibroblast in rheumatoid arthritis: cartilage destruction and the regulation of matrix metalloproteinases.* Bull NYU Hosp Jt Dis. 2006;64(1-2):20-4.

[58] Heldin CH, Westermark B. *Mechanism of action and in vivo role of platelet-derived growth factor*. Physiol Rev 1999;79:1283–316.

[59] Östman A, Heldin CH. *Involvement of platelet-derived growth factor in disease: development of specific antagonists*. Adv Cancer Res 2001;80:1–38.

[60] Rubin K, Terracio L, Ronnstrand L, Heldin CH, Klareskog L. *Expression of platelet-derived growth factor receptors is induced on connective tissue cells during chronic synovial inflammation. Scand*. J. Immunol. 1998;27:285–294.

[61] Pohlers D, Huber R, Ukena B, Kinne RW. *Expression of platelet-derived growth factors C and D in the synovial membrane of patients with rheumatoid arthritis and osteoarthritis*. Arthritis Rheum 2006;54:788–794.

[62] Thornton SC, Por SB, Penny R, Richter M, Shelley L, Breit SN. *Identification of the major fibroblast growth factors released spontaneously in inflammatory arthritis as platelet derived growth factor and tumour necrosis factor-α*. Clin Exp Immunol 1991;86:79–86.

[63] Cheon H, Sun YK, Yu SJ, Lee YH, Ji JD, Song GG, Lee JH, Kim MK, Sohn J. *Platelet derived growth factor-AA increases IL-1b and IL-8 expression and activates NF-jB in rheumatoid fibroblast-like synoviocytes*. Scand J Immunol 2004;60:455–462.

[64] Rubin K, Terracio L, Ronnstrand L, Heldin CH, Klareskog L. *Expression of platelet-derived growth factor receptors is induced on connective tissue cells during chronic synovial inflammation*. Scand J Immunol 1988;27:285–94.

[65] Kumkumian GK, Lafyatis R, Remmers EF, Case JP, Kim SJ, Wilder RL. *Platelet-derived growth factor and IL-1 interactions in rheumatoid arthritis. Regulation of synovial proliferation, prostaglandin production, and collagenase transcription*. J Immunol 1989;143:833–7.

[66] Butler DM, Leizer T, Hamilton JA. *Stimulation of human synovial fibroblast DNA synthesis by platelet-derived growth factor and fibroblast growth factor. Differences to the activation by IL-1*. J Immunol 1989;142:3098–103.

[67] Reuterdahl C, Tingström A, Terracio L, Funa K, Heldin CH, Rubin K. *Characterization of platelet-derived growth factor beta-receptor expressing cells in the vasculature of human rheumatoid synovium*. Lab Invest 1991;64:321–9.

[68] Remmers EF, Sano H, Wilder RL. *Platelet-derived growth factors and heparin-binding (fibroblast) growth factors in the synovial tissue pathology of rheumatoid arthritis*. Semin Arthritis Rheum 1991;21:191–9.

[69] Sano H, Engleka K, Mathern P, Hla T, Crofford LJ, Remmers EF, Jelsema CL, Goldmuntz E, Maciag T, Wilder RL. *Coexpression of phosphotyrosine-containing proteins, platelet-derived growth factor-B, and fibroblast growth factor-1 in situ in synovial tissues of patients with rheumatoid arthritis and Lewis rats with adjuvant or streptococcal cell wall arthritis*. J Clin Invest 1993;91:553–65.

[70] Sandler C, Joutsiniemi S, Lindstedt KA, Juutilainen T, Kovanen PT, Eklund KK. *Imatinib mesylate inhibits platelet derived growth factor stimulated proliferation of rheumatoid synovial fibroblasts*. Biochem Biophys Res Commun. 2006;347(1):31-5.

[71] George D. *Platelet-derived growth factor receptors: a therapeutic target in solid tumors*. Semin Oncol 2001;28:27–33.

[72] Nishida K, Yoshida Y, Itoh M, Fukada T, Ohtani T, Shirogane T, Atsumi T, Takahashi-Tezuka M, Ishihara K, Hibi M, Hirano T. *Gab-family adapter proteins act*

downstream of cytokine and growth factor receptors and T- and Bcell antigen receptors. Blood 1999;93:1809–16.

[73] Itoh M, Yoshida Y, Nishida K, Narimatsu M, Hibi M, Hirano T. *Role of Gab1 in heart, placenta, and skin development and growth factor- and cytokine-induced extracellular signal-regulated kinase mitogen-activated protein kinase activation.* Mol Cell Biol 2000;20:3695–704.

[74] Kallin A, Demoulin JB, Nishida K, Hirano T, Ronnstrand L, Heldin CH. *Gab1 contributes to cytoskeletal reorganization and chemotaxis in response to platelet-derived growth factor.* J Biol Chem 2004;279:17897–904.

[75] Kameda H, Ishigami H, Suzuki M, Abe T, Takeuchi T. *Imatinib mesylate inhibits proliferation of rheumatoid synovial fibroblast-like cells and phosphorylation of Gab adapter proteins activated by platelet-derived growth factor.* Clin Exp Immunol. 2006;144(2):335-41

[76] Tristano AG, Fuller K. *Immunomodulatory effects of statins and autoimmune rheumatic diseases: novel intracellular mechanism involved.* Int Immunopharmacol. 2006;6(12):1833-46.

[77] Pakozdi A, Amin MA, Haas CS, Martinez RJ, Haines GK 3rd, Santos LL, Morand EF, David JR, Koch AE. *Macrophage migration inhibitory factor: a mediator of matrix metalloproteinase-2 production in rheumatoid arthritis.* Arthritis Res Ther 2006;8(4):R132.

[78] Morel JC, Park CC, Zhu K, Kumar P, Ruth JH, Koch AE. *Signal transduction pathways involved in rheumatoid arthritis synovial fibroblast interleukin-18-induced vascular cell adhesion molecule-1 expression.* J Biol Chem. 2002;277(38):34679-91

[79] Eklund KK. *Mast cells in the pathogenesis of rheumatic diseases and as potential targets for anti-rheumatic therapy.* Immunol Rev 2007;217:38-52.

[80] Ott VL, Cambier JC, Kappler J, Marrack P, Swanson BJ. *Mast cell-dependent migration of effector CD8β T cells through production of leukotriene B4.* Nat Immunol 2003;4:974–981.

[81] Tager AM, Bromley SK, Medoff BD, Islam SA, Bercury SD, Friedrich EB, Carafone AD, Gerszten RE, Luster AD. *Leukotriene B4 receptor BLT1 mediates early effector T cell recruitment.* Nat Immunol 2003;4:982–990.

[82] Malaviya R, Twesten NJ, Ross EA, Abraham SN, Pfeifer JD. *Mast cells process bacterial Ags through a phagocytic route for to class I MHC presentation to T cells.* J Immunol 1996;156:1490–1496.

[83] Taube C, Miyahara N, Ott V, Swanson B, Takeda K, Loader J, Shultz LD, Tager AM, Luster AD, Dakhama A, Gelfand EW. *The leukotriene B4 receptor (BLT1) is required for effector CD8β T cellmediated, mast cell-dependent airway hyperresponsiveness.* J Immunol 2006;176:3157–3164.

[84] Anderson DM, Williams DE, Tushinski R, Gimpel S, Eisenman J, Cannizzaro LA, Aronson M, Croce CM, Huebner K, Cosman D. *Alternate splicing of mRNAs encoding human mast cell growth factor and localization of the gene to chromosome 12q22–q24.* Cell Growth Differ 1991;2:373–378.

[85] Kiener HP, Hofbauer R, Tohidast-Akrad M, Walchshofer S, Redlich K, Bitzan P, Kapiotis S, Steiner G, Smolen JS, Valent P. *Tumor necrosis factor alpha promotes the expression of stem cell factor in synovial fibroblasts and their capacity to induce mast cell chemotaxis.* Arthritis Rheum 2000;43:164–174.

[86] Nilsson G, Butterfield JH, Nilsson K, Siegbahn A. *Stem cell factor is a chemotactic factor for human mast cells.* J Immunol 1994;153:3717–3723.
[87] Reber L, Da Silva CA, Frossard N. *Stem cell factor and its receptor c-Kit as targets for inflammatory diseases.* Eur J Pharmacol 2006;533:327–340.
[88] Reber L, Da Silva CA, Frossard N. *Stem cell factor and its receptor c-Kit as targets for inflammatory diseases.* Eur J Pharmacol. 2006;533(1-3):327-40.
[89] Yarden Y, Kuang WJ, Yang-Feng T, Coussens L, Munemitsu S, Dull TJ, Chen E, Schlessinger J, Francke U, Ullrich A. *Human proto-oncogene c-kit: a new cell surface receptor tyrosine kinase for an unidentified ligand.* EMBO J 1987;6:3341–3351
[90] Buzby JS, Knoppel EM, Cairo MS. *Coordinate regulation of steel factor; its receptor (Kit), and cytoadhesion molecule (ICAM-1 and ELAM-1) mRNA expression in human vascular endothelial cells of differing origins.* Exp Hematol. 1994;22:122–129.
[91] Sillaber C, Strobl H, Bevec D, Ashman LK, Butterfield JH, Lechner K, Maurer D, Bettelheim P, Valent P. *IL-4 regulates c-kit protooncogene product expression in human mast cell and myeloid progenitor cells.* J Immunol. 1991;147:4224–4228.
[92] Welker P, Grabbe J, Zuberbier T, Grutzkau A, Henz BM. *GM-CSF downmodulates c-kit, Fc(epsilon)RI(alpha) and GM-CSF receptor expression as well as histamine and tryptase levels in cultured mast cells.* Arch Dermatol Res. 2001;293:249–258.
[93] Mirmonsef P, Shelburne CP, Fitzhugh Yeatman II C, Chong HJ, Ryan JJ. *Inhibition of kit expression by IL-4 and IL-10 in murine mast cells: role of STAT6 and phosphatidylinositol 3'-kinase.* J Immunol 1999;163:2530–2539.
[94] Haraoui B. *The anti-tumor necrosis factor agents are a major advance in the treatment of rheumatoid arthritis.* J. Rheumatol Suppl. 2005;72:46–47.
[95] Kinne RW, Bräuer R, Stuhlmüller B, Palombo-Kinne E, Burmester GR. *Macrophages in rheumatoid arthritis.* Arthritis Res. 2000;2(3):189-202.
[96] Pixley FJ, Stanley ER. *CSF-1 regulation of the wandering macrophage: complexity in action.* Trends Cell Biol. 2004;14(11):628-38.
[97] Campbell IK, Rich MJ, Bischof RJ, Hamilton JA. *The colony-stimulating factors and collagen-induced arthritis: exacerbation of disease by M-CSF and G-CSF and requirement for endogenous M-CSF.* J Leukoc Biol 2000;68:144–150.
[98] Mustonen T, Alitalo K. *Endothelial receptor tyrosine kinases involved in angiogenesis.* J Cell Biol. 1995;129:895-898.
[99] Ferrara N, Davis-Smyth T. *The biology of vascular endothelial growth factor.* Endocr Rev. 1997;18:4-25.
[100] Risau W. *Mechanisms of angiogenesis.* Nature. 1997;386:671-674.
[101] Shibuya M. *Structure and function of VEGF/VEGF-receptor system involved in angiogenesis.* Cell Struct Funct. 2001;26:25-35.
[102] Shibuya M, Yamaguchi S, Yamane A, Ikeda T, Tojo A, Matsushime H, Sato M. *Nucleotide sequence and expression of a novel human receptor-type tyrosine kinase gene (flt) closely related to the fms family.* Oncogene 1990;5:519-524.
[103] Kondo K, Hiratsuka S, Subbalakshmi E, Matsushime H, Shibuya M. *Genomic organization of the flt-1 gene encoding for vascular endothelial growth factor (VEGF) receptor-1 suggests an intimate evolutionary relationship between the 7-Ig and the 5-Ig tyrosine kinase receptors.* Gene 1998;208:297-305.

[104] Jakeman LB, Winer J, Bennett GL, Altar CA, Ferrara N. *Binding sites for vascular endothelial growth factor are localized on endothelial cells in adult rat tissues.* J Clin Invest. 1992;89:244-253.

[105] Eichmann A, Marcelle C, Breant C, Le Douarin NM. *Two molecules related to the VEGF receptor are expressed in early endothelial cells during avian embryonic development.* Mech Dev. 1993;42:33-48.

[106] Kaipainen A, Korhonen J, Pajusola K, Aprelikova O, Persico MG, Terman BI, Alitalo K. *The related FLT4, FLT1, and KDR receptor tyrosine kinases show distinct expression patterns in human fetal endothelial cells.* J Exp Med. 1993;178:2077-2088.

[107] Yamane A, Seetharam L, Yamaguchi S, Gotoh N, Takahashi T, Neufeld G, Shibuya M. *A new communication system between hepatocytes and sinusoidal endothelial cells in liver through vascular endothelial growth factor and Flt tyrosine kinase receptor family (Flt-1 and KDR/Flk-1).* Oncogene. 1994;9:2683-2690.

[108] Barleon B, Sozzani S, Zhou D, Weich HA, Mantovani A, Marme D. *Migration of human monocytes in response to vascular endothelial growth factor (VEGF) is mediated via the VEGF receptor flt-1.* Blood. 1996;87:3336-3343.

[109] Clauss M, Weich H, Breier G, Knies U, Röckl W, Waltenberger J, Risau W. *The vascular endothelial growth factor receptor Flt-1 mediates biological activities: implications for a functional role of placenta growth factor in monocyte activation and chemotaxis.* J Biol Chem. 1996;271:17629-17634.

[110] Sawano A, Iwai S, Sakurai Y, Ito M, Shitara K, Nakahata T, Shibuya M. *Flt-1, vascular endothelial growth factor receptor 1, is a novel cell surface marker for the lineage of monocytemacrophages in humans.* Blood. 2001;97:785-791.

[111] Fava RA, Olsen NJ, Spencer-Green G, Yeo KT, Yeo TK, Berse B, Jackman RW, Senger DR, Dvorak HF, Brown LF. *Vascular permeability factor/endothelial growth factor (VPF/VEGF): accumulation and expression in human synovial fluids and rheumatoid synovial tissue.* J Exp Med. 1994;180:341-346.

[112] Koch AE, Harlow LA, Haines GK, Amento EP, Unemori EN, Wong WL, Pope RM, Ferrara N. *Vascular endothelial growth factor: a cytokine modulating endothelial function in rheumatoid arthritis.* J Immunol. 1994;152:4149-4156.

[113] Murakami M, Iwai S, Hiratsuka S, Yamauchi M, Nakamura K, Iwakura Y, Shibuya M. *Signaling of vascular endothelial growth factor receptor-1 tyrosine kinase promotes rheumatoid arthritis through activation of monocytes/macrophages.* Blood. 2006;108(6):1849-56.

[114] De Bandt M, Ben Mahdi MH, Ollivier V, Grossin M, Dupuis M, Gaudry M, Bohlen P, Lipson KE, Rice A, Wu Y, Gougerot-Pocidalo MA, Pasquier C. *Blockade of vascular endothelial growth factor receptor I (VEGF-RI), but not VEGF-RII, suppresses joint destruction in the K/BxN model of rheumatoid arthritis.* J Immunol. 2003;171:4853-4859.

[115] Giatromanolaki A, Sivridis E, Athanassou N, Zois E, Thorpe PE, Brekken RA, Gatter KC, Harris AL, Koukourakis IM, Koukourakis MI. *The angiogenic pathway 'vascular endothelial growth factor/flk-1(KDR)-receptor' in rheumatoid arthritis and osteoarthritis.* J Pathol 2001;194:101–8.

[116] Grosios K, Wood J, Esser R, Raychaudhuri A, Dawson J. *Angiogenesis inhibition by the novel VEGF receptor tyrosine kinase inhibitor, PTK787/ZK222584, causes*

significant anti-arthritic effects in models of rheumatoid arthritis. Inflamm Res. 2004;53(4):133-42.

[117] Suda T, Nakamura I, Jimi E, Takahashi N. *Regulation of osteoclast function.* J Bone Miner Res 1997;12:869–79.

[118] Miyazaki T, Tanaka S, Sanjay A, Baron R. *The role of c-Src kinase in the regulation of osteoclast function.* Mod Rheumatol. 2006;16(2):68-74.

[119] Miyazaki T, Takayanagi H, Isshiki M, Takahashi T, Okada M, Fukui Y, Oda H, Nakamura K, Hirai H, Kurokawa T, Tanaka S. *In vitro and in vivo suppression of osteoclast function by adenovirus vector-induced csk gene.* J Bone Miner Res 2000;15:41–51.

[120] Soriano P, Montgomery C, Geske R, Bradley A. *Targeted disruption of the c-src proto-oncogene leads to osteopetrosis in mice.* Cell 1991;64:693–702.

[121] Miyazaki T, Sanjay A, Neff L, Tanaka S, Horne WC, Baron R. *Src kinase activity is essential for osteoclast function.* J Biol Chem 2004;279:17660 -6.

[122] Takayanagi H, Juji T, Miyazaki T, Iizuka H, Takahashi T, Isshiki M, Okada M, Tanaka Y, Koshihara Y, Oda H, Kurokawa T, Nakamura K, Tanaka S. *Suppression of arthritic bone destruction by adenovirus-mediated csk gene transfer to synoviocytes and osteoclasts.* J Clin Invest. 1999;104(2):137-46.

[123] Nada S, Okada M, MacAuley A, Cooper JA, Nakagawa H. *Cloning of a complementary DNA for a protein-tyrosine kinase that specifically phosphorylates a negative regulatory site of p60c-src.* Nature. 1991;351:69–72.

[124] Okada M, Nada S, Yamanashi Y, Yamamoto T, Nakagawa H. *CSK: a protein-tyrosine kinase involved in regulation of src family kinases.* J Biol Chem. 1991;266:24249–24252.

[125] Dewar AL, Zannettino AC, Hughes TP, Lyons AB. *Inhibition of c-fms by imatinib: expanding the spectrum of treatment.* Cell Cycle. 2005;4(7):851-3

[126] Cope AP, Schulze-Koops H, Aringer M. *The central role of T cells in rheumatoid arthritis.* Clin Exp Rheumatol. 2007;25(5 Suppl 46):S4-11

[127] van den Berg WB, van Lent PL, Joosten LA, Abdollahi-Roodsaz S, Koenders MI. *Amplifying elements of arthritis and joint destruction.* Ann Rheum Dis. 2007;66 Suppl 3:iii45-8.

[128] Fournier C. *Where Do T Cells Stand in Rheumatoid Arthritis?* Joint Bone Spine 2005;72:527–532.

[129] Nogid A, Pham DQ. *Role of abatacept in the management of rheumatoid arthritis.* Clin Ther. 2006;28(11):1764-78.

[130] Qian D, Weiss A. *T cell antigen receptor signal transduction.* Curr Opin Cell Biol. 1997;9:205-212.

[131] van Oers NS, Killeen N, Weiss A. *Lck regulates the tyrosine phosphorylation of the T cell receptor subunits and ZAP-70 in murine thymocytes.* J Exp Med. 1996;183:1053-1062.

[132] Leder C, Ortler S, Seggewiss R, Einsele H, Wiendl H. *Modulation of T-effector function by imatinib at the level of cytokine secretion.* Exp Hematol. 2007;35(8):1266-71

[133] Feldmann M, Brennan FM, Maini RN. *Role of cytokines in rheumatoid arthritis.* Annu Rev Immunol 1996;14:397–440.

[134] Kusada J, Otsuka T, Matsui N, Hirano T, Asai K, Kato T. *Immuno-reactive human epidermal growth factor (h-EGF) in rheumatoid synovial fluids.* Nippon Seikeigeka Gakkai Zasshi 1993;67:859–65.

[135] Wen D, Peles E, Cupples R, Suggs SV, Bacus SS, Luo Y, Trail G, Hu S, Silbiger SM, Levy RB. *Neu differentiation factor: a transmembrane glycoprotein containing an EGF domain and an immunoglobulin homology unit.* Cell 1992;69:559–72.

[136] Carraway KI III, Cantley LC. *A neu acquaintance for erbB3 and erbB4: a role for receptor heterodimerization in growth signaling.* Cell 1994;78:5–8.

[137] Holmes WE, Sliwkowski MX, Akita RW, Henzel WJ, Lee J, Park JW, Yansura D, Abadi N, Raab H, Lewis GD . *Identification of heregulin, a specific activator of p185erbB2.* Science 1992;256:1205–10.

[138] Riese DJ II, van Raaij TM, Plowman GD, Andrews GC, Stern DF. *The cellular response to neuregulins is governed by complex interactions of the erbB receptor family.* Mol Cell Biol 1995;15:5770–6.

[139] Satoh K, Kikuchi S, Sekimata M, Kabuyama Y, Homma MK, Homma Y. *Involvement of ErbB-2 in rheumatoid synovial cell growth.* Arthritis Rheum. 2001;44(2):260-5.

[140] Ivashkiv LB. *Cytokine expression and cell activation in inflammatory arthritis.* Adv Immunol 1996;63:337–76.

[141] Leonard WJ, O'Shea JJ. *Jaks and STATs: biological implications.* Annu Rev Immunol 1998;16:293–322.

[142] Migita K, Komori A, Torigoshi T, Maeda Y, Izumi Y, Jiuchi Y, Miyashita T, Nakamura M, Motokawa S, Ishibashi H. *CP690,550 inhibits oncostatin M-induced JAK/STAT signaling pathway in rheumatoid synoviocytes.* Arthritis Res Ther. 2011;13(3):R72

[143] Meyer DM, Jesson MI, Li X, Elrick MM, Funckes-Shippy CL, Warner JD, Gross CJ, Dowty ME, Ramaiah SK, Hirsch JL, Saabye MJ, Barks JL, Kishore N, Morris DL. *Anti-inflammatory activity and neutrophil reductions mediated by the JAK1/JAK3 inhibitor, CP-690,550, in rat adjuvant-induced arthritis.* J Inflamm (Lond). 2010;7:41

[144] Fridman JS, Scherle PA, Collins R, Burn TC, Li Y, Li J, Covington MB, Thomas B, Collier P, Favata MF, Wen X, Shi J, McGee R, Haley PJ, Shepard S, Rodgers JD, Yeleswaram S, Hollis G, Newton RC, Metcalf B, Friedman SM, Vaddi K. *Selective inhibition of JAK1 and JAK2 is efficacious in rodent models of arthritis: preclinical characterization of INCB028050.* J Immunol. 2010;184(9):5298-307

[145] DeBusk LM, Chen Y, Nishishita T, Chen J, Thomas JW, Lin PC. *Tie2 receptor tyrosine kinase, a major mediator of tumor necrosis factor alpha-induced angiogenesis in rheumatoid arthritis.* Arthritis Rheum. 2003;48(9):2461-71.

[146] Dumont DJ, Gradwohl G, Fong GH, Puri MC, Gertsenstein M, Auerbach A, Breitman ML. *Dominant-negative and targeted null mutations in the endothelial receptor tyrosine kinase, tek, reveal a critical role in vasculogenesis of the embryo.* Genes Dev 1994;8:1897–909.

[147] Sato TN, Qin Y, Kozak CA, Audus KL. *Tie-1 and tie-2 define another class of putative receptor tyrosine kinase genes expressed in early embryonic vascular system.* Proc Natl Acad Sci U S A 1993;90:9355–8.

[148] Davis S, Aldrich TH, Jones PF, Acheson A, Compton DL, Jain V, Ryan TE, Bruno J, Radziejewski C, Maisonpierre PC, Yancopoulos GD. *Isolation of angiopoietin-1, a*

ligand for the TIE2 receptor, by secretion-trap expression cloning. Cell 1996;87:1161–9.

[149] Uchida T, Nakashima M, Hirota Y, Miyazaki Y, Tsukazaki T, Shindo H. *Immunohistochemical localisation of protein tyrosine kinase receptors Tie-1 and Tie-2 in synovial tissue of rheumatoid arthritis: correlation with angiogenesis and synovial proliferation.* Ann Rheum Dis 2000;59:607–14.

[150] Scola MP, Imagawa T, Boivin GP, Giannini EH, Glass DN, Hirsch R, Grom AA. *Expression of angiogenic factors in juvenile rheumatoid arthritis: correlation with revascularization of human synovium engrafted into SCID mice.* Arthritis Rheum 2001;44:794–801.

[151] Pine PR, Chang B, Schoettler N, Banquerigo ML, Wang S, Lau A, Zhao F, Grossbard EB, Payan DG, Brahn E. *Inflammation and bone erosion are suppressed in models of rheumatoid arthritis following treatment with a nove Syk inhibitor.* Clin Immunol. 2007;124(3):244-57.

[152] Berton G, Mocsai A, Lowell CA. *Src and Syk kinases: key regulators of phagocytic cell activation.* Trends Immunol. 2005;26:208–214.

[153] Kyttaris VC, Tsokos GC. *Syk kinase as a treatment target for therapy in autoimmune diseases.* Clin Immunol. 2007;124(3):235-7.

[154] Cha HS, Boyle DL, Inoue T, Schoot R, Tak PP, Pine P, Firestein GS. *The novel Syk inhibitor R406 blocks JNK-mediated gene expression in synoviocytes,* J. Pharmacol. Exp Ther. 2006;317:571–578.

[155] Lindqvist E, Eberhardt K, Bendtzen K, Heinegard D, Saxne T. *Prognostic laboratory markers of joint damage in rheumatoid arthritis.* Ann Rheum Dis. 2005;64(2):196–201.

[156] Larsson E, Erlandsson Harris H, Lorentzen JC, Larsson A, Månsson B, Klareskog L, Saxne T. *Serum concentrations of cartilage oligomeric matrix protein, fibrinogen and hyaluronan distinguish inflammation and cartilage destruction in experimental arthritis in rats.* Rheumatology 2002;41(9):996–1000.

[157] Weinblatt ME, Kavanaugh A, Burgos-Vargas R, Dikranian AH, Medrano-Ramirez G, Morales-Torres JL, Murphy FT, Musser TK, Straniero N, Vicente-Gonzales AV, Grossbard E. *Treatment of rheumatoid arthritis with a Syk kinase inhibitor: a randomized twelve-week, randomized, placebo- controlled trial.* Arthritis Rheum 2008;58:3309-18.

[158] Weinblatt ME, Kavanaugh A, Genovese MC, Musser TK, Grossbard EB, Magilavy DB. *An oral spleen tyrosine kinase (Syk) inhibitor for rheumatoid arthritis.* N Engl J Med. 2010;363(14):1303-12

[159] Genovese MC, Kavanaugh A, Weinblatt ME, Peterfy C, DiCarlo J, White ML, O'Brien M, Grossbard EB, Magilavy DB. *An oral Syk kinase inhibitor in the treatment of rheumatoid arthritis: a three-month randomized, placebo-controlled, phase II study in patients with active rheumatoid arthritis that did not respond to biologic agents.* Arthritis Rheum. 2011;63(2):337-45.

[160] Kremer JM, Bloom BJ, Breedveld FC, Coombs JH, Fletcher MP, Gruben D, Krishnaswami S, Burgos-Vargas R, Wilkinson B, Zerbini CA, Zwillich SH. *The safety* Arthritis Rheum. 2009;60(7):1895-905.

[161] Coombs JH, Bloom BJ, Breedveld FC, Fletcher MP, Gruben D, Kremer JM, Burgos-Vargas R, Wilkinson B, Zerbini CA, Zwillich SH. *Improved pain, physical functioning*

and health status in patients with rheumatoid arthritis treated with CP-690,550, an orally active Janus kinase (JAK) inhibitor: results from a randomised, double-blind, placebo-controlled trial. Ann Rheum Dis. 2010;69(2):413-6.

[162] Cohen S, Zwillich SH, Chow V, Labadie RR, Wilkinson B. *Co-administration of the JAK inhibitor CP-690,550 and methotrexate is well tolerated in patients with rheumatoid arthritis without need for dose adjustment.* Br J Clin Pharmacol. 2010;69(2):143-51.

[163] Kremer JM, Cohen S, Wilkinson BE, Connell CA, French JL, Gomez-Reino J, Gruben D, Kanik KS, Krishnaswami S, Pascual-Ramos V, Wallenstein G, Zwillich SH, *A Phase 2B dose-ranging study of the oral JAK inhibitor tofacitinib (CP-690,550) versus placebo in combination with background methotrexate in patients with active rheumatoid arthritis and inadequate response to methotrexate alone.* Arthritis Rheum. 2011. Doi: 10.1002/art 33419.

[164] Akashi N, Matsumoto I, Tanaka Y, Inoue A, Yamamoto K, Umeda N, Tanaka Y, Hayashi T, Goto D, Ito S, Sekiguchi K, Sumida T. *Comparative suppressive effects of tyrosine kinase inhibitors imatinib and nilotinib in models of autoimmune arthritis.* Mod Rheumatol. 2011;21(3):267-75.

[165] Milici AJ, Kudlacz EM, Audoly L, Zwillich S, Changelian P. *Cartilage preservation by inhibition of Janus kinase 3 in two rodent models of rheumatoid arthritis.* Arthritis Res Ther. 2008;10(1):R14.

[166] Stump KL, Lu LD, Dobrzanski P, Serdikoff C, Gingrich DE, Dugan BJ, Angeles TS, Albom MS, Ator MA, Dorsey BD, Ruggeri BA, Seavey MM. *A highly selective, orally active inhibitor of Janus kinase 2, CEP-33779, ablates disease in two mouse models of rheumatoid arthritis.* Arthritis Res Ther. 2011;13(2):R68.

[167] Tristano AG. *Tyrosine* Int Immunopharmacol. 2009;9(1):1-9.

In: Tyrosine and Aspartic Acid
Editors: J. E. Jones and D. M. Morano
ISBN: 978-1-62100-752-4

Chapter 2

TYROSINES IN BIOLOGICAL ELECTRON TRANSFER

Mariana Voicescu[1,2] and Petra Hellwig[1]
[1]Laboratoire de Spectroscopie Vibrationnelle et Electrochimie des Biomolécules, UMR 7177, Institut de Chimie, CNRS-Université de Strasbourg, 1 rue Blaise Pascal, Strasbourg, France
[2]Romanian Academy, Institute of Physical Chemistry "Ilie Murgulescu", Splaiul Independentei 202, Bucharest, Romania

ABSTRACT

Specific amino acid side chains are crucial to the stability of proteins structure and in the catalysis of enzymatic reactions. An important role is granted to Tyr residues and Tyr derivatives originating from posttranslational modifications. An example for such a modified Tyr is found in cytochrome *c* oxidase, where a covalent linkage between a Tyr and a His was found. The role of this linkage for the catalytic mechanism was explored by a number of research groups, several of the studies including model compound studies. Electrochemical and spectroscopic studies gave evidence to the influence of this linkage on the electron transfer ability of the Tyr and its function within the catalytic mechanism.

1. INTRODUCTION

1.1. General Aspects on Proteins in Biological Processes

Proteins are at the center of action in biological processes and all the molecular transformations that define cellular metabolism are mediated by protein catalysis [1]. Life forms make use of many chemical reactions to supply themselves continually with chemical energy and to use it efficiently, but these reactions could not occur fast enough under physiological conditions (aqueous solution, 37 °C and pH 7, atmospheric pressure) to sustain life. The rates of these reactions are increased, by the presence of proteins, so called enzymes, by many orders of magnitude [2]. In spite of all these diverse biological functions, proteins are a relatively homogeneous class of molecules and are the same type of linear polymers, built of various combinations of the same 20 amino acids. They differ in the sequence in which amino acids are assembled into polymeric chains. The secret to their functional diversity lies partly in the chemical diversity of the amino acid but primarily in the diversity of the three-dimensional structures that these building blocks can form, simply by being linked in different sequences [3] and in their secondary, tertiary and quartary structures [4]. Individual amino acids are crucial to assembly, stability or function, while some of the amino acids do not participate to the activity or stability of a protein.

1.2. The Redox Activity of the Proteins

Redox reactions are the basis of the energy delivering processes in living organisms, including respiration and photosynthesis.

Most redox active proteins contain a metal center (metalloproteins) [5]; However, also proteins not including a metal center can show redox activity. It was reported that peptides and proteins containing cysteine/cystine (Cys) can show specific electrochemical signals on mercury electrodes after Hg-S bond formation [6, 7], the reduction of disulfide groups [8], and the catalytic evolution of hydrogen in cobalt containing solutions (Brdicka reaction). The later is mostly known for proteins that contain sulfur based amino acids [9]. A significant class of redox transfer reactions involve the electron transfer via aromatic residues like quinones, flavins [10, 11] and posttranslationally modificed cofactors, like 2, 4, 5-trihydroxyphenylalanine quinone (TPQ or TOPA quinone), pyrroloquinoline quinone (Ref [12] and references therein) where Tyrosine residues were found to be extensively used in nature to generate proteinaceous cofactors.

1.2.1. The Specific Role of the Redox Couple Tyrosine/Tyrosine Radical

Reynaud et *al.* [13] and Brabec et *al.* [14, 15] showed that Tyrosine (Tyr) and Tryptophan (Trp) residues are the main residues in redox-active proteins. The redox reaction of Tyr residues involves the transfer of two electrons and two protons, the electrode process being similar to the oxidation of simple p-substituted phenols [13-16]. For label-free detection of amino acids and peptides, which is based on the inherent electrochemical properties of the analyte by using direct or mediated (in solution or immobilised) redox processes and interfacial ion transfers, the oxidation of Tyr and Trp are in the concentration range of some nM [17, 18]. This feature leads to the use of redox-active amino acids residues for protein sensing.

The phenolic amino acid L-Tyr is present in human blood plasma; The one-electron oxidation of L-Tyr by potassium permanganate generates tyrosyl radicals [19]. When two tyrosyl radicals productively interact, the major product is *o*, *o*'-diTyr (3, 3'-diTyr), an intensely fluorescent di-amino acid [20, 21]. The reaction pathway bears remarkable similarities to the chemical generation of phenoxyl radical [22, 23]. In both systems, a one-electron oxidation of a phenol yields a phenoxyl radical, which can form a wide variety of carbon-carbon and carbon-oxygen cross-linked products. Also, peroxidases can oxidize Tyr to *o*, *o*'-diTyr *in vitro* [24, 25]. The reaction also occurs in vivo, for example when the ovo-peroxidase of nascent sea urchin embryos generates *o*, *o*'-diTyr cross-links in the protective fertilization envelope by a reaction requiring H_2O_2 [26]. A cytochrome P450-like enzyme similarly cross-links proteins in the spore wall of the yeast *Sacchraromyces* [27, 28]; It is not well understood whether these reactions involve tyrosyl radical and whether myeloperoxidase of activated human phagocytes also produces *o, o'*-diTyr cross-links.

Because the interfacial behavior of proteins plays an important role in understanding the mechanisms involved in the course of biochemical processes and subsequently in the medical problems associated with the fouling of surfaces by proteins the adsorption and electrochemistry of some essential amino acids was in focus [29-31]. Particular attention was given to the study of amino acids containing aromatic groups, such as phenylalanine and Tyr, since they exhibit biochemical functionalities in the regulation of the growth, survival and repair of the nervous system [32]. Tyrosine, among the free other aromatic aminoacid residues, has been found to be electro-active [15], an important fact in the cleavage process of dipeptides.

The protein environment of the Tyr residue can control the function of the redox active aminoacid. The aromatic side chain of Tyr is involved in long-distance electron-transfer reactions in enzymes, reactions that are essential steps in the enzyme's catalytic mechanism. The redox couple Tyrosyl radical/Tyrosine, has been observed in key enzymes such as ribonucleotide reductase [33-35], which provide precursors of DNA synthesis, and also in photosystem II [36, 37], the membrane protein complex where photosynthetic oxygen evolution takes place in plants, algae and cyanobacteria. Stable and transient Tyr radicals are involved in a variety of regulatory enzymes that have an important role in primary metabolism. Prostaglandin synthases, essential for the production of a variety of prostaglandins from arachidonic acid, involve both a cyclooxygenase reaction, which converts arachidonic acid to prostaglandin G_2, and the hydroxylation of prostaglandin G_2 into prostaglandin H_2. Furthermore, the peroxidation reaction involves a highly oxidizing intermediate, an oxoferryl protoporphyrin IX radical cation [38, 39] that generates a Tyr radical by intraprotein electron transfer. The Tyr radical in turn initiates the cyclooxygenase reaction by forming the arachidonate radical [40, 41] which reacts with oxygen to yield prostaglandin G_2. Crystallographic data [42] showed the location of a Tyr radical between the heme and the arachidonate binding sites, which couples the reaction of the two enzymes [42]. Catalases and peroxidases have a protoporphyrin prosthetic group at the active site or, alternately, an oxoferryl protoporphyrin IX and a protein based radical species that are generated by the two-electron oxidation of the protoporphyrin by peroxides [43].

Great interest was conferred to DNA photolyase, flavoproteins that catalyze UV-induced repair of major UV-damaged DNA lesions by reduction of pyrimidine dimmers, and cryptochrome-1; An electron transfer from a Tyr residue to a Trp radical was detected in the DNA photolyase of *Anacystis nidulans* [44]. Furthermore, a transient Tyr radical was detected in the blue light photoreceptor cryptochrome-1 from *Arabidopsis thaliana* [45] that presents a significant sequence analogy with DNA-photolyase

1.2.2. Modified Redox Active Tyrosines (Posttranslational)

Modified Redox active Tyrosines are involved in the mechanism of a number of metallo-radical enzymes that couple substrate oxidation with oxygen reduction to hydrogen peroxide or water.

Thioether-Tyr, resulting from the covalent binding of the sulfur atom of adjacent cysteine at the α- position of the phenol ring, was identified at the active site of galactose oxidase [46-

48] and glyoxal oxidase [49]. Glyoxal oxidase is a fungal enzyme and the physiological partner of lignin peroxidase and Mn peroxidase. It couples the oxidation of a number of aldehydes and a-hydroxy carbonyl compounds with the reduction of O_2 to H_2O_2. Galactose oxidase catalyzes the two-electron oxidation of primary alcohols to aldehydes with the concomitant reduction of dioxygen to hydrogen peroxide. Although dissimilar in primary sequence, the two enzymes have very similar active sites, in which the modified Tyr directly coordinates to the active site copper ion [50]. Catalysis involves redox reactions of both the copper and cysteinyl-Tyr ligand (Ref. [51] and references therein). In galactose oxidase, oxidation of the substrate alcohol involves a two-electron oxidation, deprotonation and the cleavage of one of the C_α–H bonds within the hydroxymethylene group. This occurs through oxidation of the copper, protonation of the Tyrosinate Cu ligand and cleavage of the C_α–H bond by the cysteinyl-Tyr ligand. The strong kinetic isotope effect associated with the oxidation of C_α-^{1}H or C_α-^{2}H-labeled substrates and the strong temperature dependence of this effect were interpreted in favor of a mechanism based on a hydrogen atom abstraction from the substrates by the coordinated cysteinyl - Tyr radical [51, 52]. For the prostaglandin H-synthase, a direct involvement in the oxidation of the substrate arachidonic acid [38], of Mn - bound water, or hydroxyl [52-55] was reported through proton and electron abstraction from C–H or O–H bonds.

In copper amine oxidase (2, 4, 5-trihydroxyphenylalanine quinone), topaquinone (TPQ) cofactor results from a posttranslational copper-dependent oxidation of a conserved Tyr residue; During the enzyme turnover, TPQ forms a transient iminosemiquinone radical [56, 57].

In cytochrome *c* oxidase, the terminal enzyme in the respiratory chain of the mitochondria, as well as in other heme-copper terminal oxidases, the active site for oxygen reduction is located at a binuclear metal center formed by a high-spin heme A (heme a_3) and copper Cu_B [58-61]. A Tyr residue forms a covalent bond to His ligand of Cu_B [58, 59], and it was found to be redox-active by $^{125}I^-$ labeling and peptide mapping analysis [62]. Tyr-His radical was detected during the reaction of the heme copper oxidases with hydrogen peroxide [63, 64].

1.2.3. The Role of the Posttranslationally Modified Tyr on the Cco Mechanism

Cytochrome *c* Oxidase (C*c*O) catalyzes the reduction of dioxygen to water [60, 65]. Water formation requires the translocation of four protons. Additional four protons are pumped across the membrane contributing to the electrochemical proton gradient, in producing proton motive force, which is used in ATP synthesis. Electron injection from cytochrome *c* to the homo- dinuclear copper center, CuA (electron transfer), is followed by intramolecular electron transfer, via the low-spin heme *a* (electron transfer), to the binuclear center which contains a high spin heme a_3 and a Cu_B atom. The heme a_3-Cu_B center, where dioxygen reduction occurs, represents a crucial site for proton gating as a proton translocation has been shown to couple to the steps in the oxygen reduction pathway [66-72]. X-ray crystallographic [73,74] and biochemical [75, 76] studies reveal that one of the copper-bound histidines is covalently linked to a Tyrosine through a posttranslationally modified cross-linked Histidine between C6 of Tyr^{244}/Tyr^{280} and the ε-nitrogen of His^{240}/His^{276}(*Bovine* and *Paracoccus denitrificans* numbering, respectively). Figure 1 shows the active site of the cytochrome *c* oxidase, that contains a binuclear center composed from heme a_3 and Cu_B and the Tyr-His covalent linkage (*Paracoccus denitrificans* numbering). This linkage was suggested to be created during the first turnover of the enzyme by the oxygen reaction intermediates.

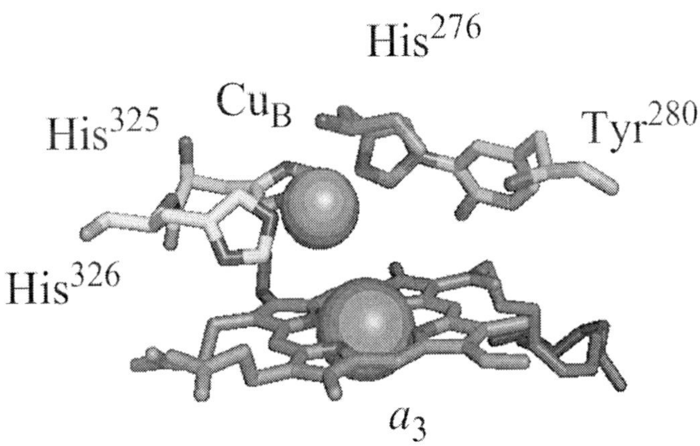

http://www.pymol.org.

Figure 1. The active site and the Tyr-His crosslink of the cytochrome *c* oxidase from *Paracoccus denitrificans*. The figure was prepared using the 1M56 PDB file and the PyMol software.

While His^{240} is one of the ligands of Cu_B, Tyr^{244} seems to be a possible redox center in which it holds an extra oxidation equivalent in the **P** intermediate of its enzymatic reaction [59, 62, 77, 78]. The proton transfer from Tyr^{244} to a ferric peroxide generates a hydroperoxo adduct and subsequently Cu_B was suggested to provide an electron via His^{240}-Tyr^{244} cross-link to cleave the O-O bond of the ferric hydroperoxide; The cross-linked Tyr could serve as a proton donor during the O_2 reduction [58, 69-71, 79]. The metals in the active site provide three of the four electrons required for O-O bond cleavage by oxidation of the Fe^{2+}/Cu_B^{+} binuclear center to oxoferryl (Fe^{4+}= O) and HO-Cu_B^{2+}. Within these lines, it is discussed that in the cross-linked Tyr-His, the Tyr residue might supply both the proton and the fourth electron required for the transformation via hydrogen atom transfer from the neutral Tyr. A Tyrosyl radical would be generated, placing C*c*O in the company of a growing number of proteins that utilize redox active amino acids during catalysis [33]. It was shown that the ability to transfer electron and protons will dependent on the physicochemical properties of the Tyr that are changed as a result of the covalent bonding to His [59].

Different protonation states are suggested for the reaction of this specific Tyr residue [80-84], depending on the pH and other experimental conditions. Whereas a deprotonation of the residue before electron transfer is necessary, it is under debate at which part of the cycle the tyrosine deprotonates and subsequently participates in the redox reaction. The direct observation of protonation reactions during the catalytic cycle of C*c*O was reported by Nyquist et *al* [80] and the deprotonation of a redox-active site Tyr, plausibly Tyr^{280}, in the **F** intermediate, is suggested. Structural and chemical changes of the $\mathbf{P_M}$ from *Paracoccus denitrificans* C*c*O were investigated by Iwaki et *al* [81, 82]. In these experiments transitions from the oxidized to the $\mathbf{P_M}$ state were initiated by perfusion with CO/oxygen buffer. The investigation of Tyr-ring-d^4 labeled protein had large effects on $\mathbf{P_M}$ minus oxidized difference spectra, leading to the attribution of a protonated Tyr hydroxyl group present for the oxidized state of the binuclear center. In addition the deprotonation of a Tyr residue with the formation of the $\mathbf{P_M}$ intermediate was presented [81]. By labeling both the Tyr- and the His-group, the isotope effects on the amide I regions in $\mathbf{P_M}$ minus **O** spectra, it was suggested that the amide carbonyl bonds of Tyr and His are major contributors and subsequently, a structural alteration in the $\mathbf{P_M}$ intermediate was suggested [82]. Using the CO photolysis approach for reaction intermediate formation, the protonation state of Tyr in all the main intermediates throughout the catalytic cycle is reported by Gorbikova et *al.* [83], Figure 2. It was also suggested that the deprotonated Tyr^{280} must be reprotonated later on in the catalytic cycle to serve as a proton donor for the next oxygen reduction event. Moreover, the complete reprotonation of the Tyr is linked to the formation of the one-electron reduced state coupled to the reduction of the Cu_B site [83]. Also, the development of electrochemical potential during formation of the $\mathbf{P_R}$ species suggests transfer of a proton oriented perpendicularly, to the membrane plane and over a distance of 4 Å, which is close to the distance between the oxygen atom of the hydroxyl group of Tyr^{280} and the bound oxygen [84].

2. Tyr-His Model Compounds

2.1. Tyr and Tyr-His Crosslink Importance in the Functional Properties of the Proteins

2.1.1. The Molecular Structure of the Tyr - His Model Compounds

As mentioned above, Tyr radicals play a crucial role during the catalytic reaction in redox enzymes [46-51]. For a clearer understanding of the influence of the covalent linkage on the physicochemical properties of Tyrosines and thus on the formation of tyrosine radicals in general, detailed spectroscopic and electrochemical studies on models of these cross-linked molecules have been made. Studies leading to insights into the functional role of the Tyr-His linkage in C*c*O were reported [85-91]. Figure 3 shows the molecular structure of such cross-linked Tyr-His model compounds together with His-Tyr Cu (II) complexes, synthesized by different research groups [85, 92-95].

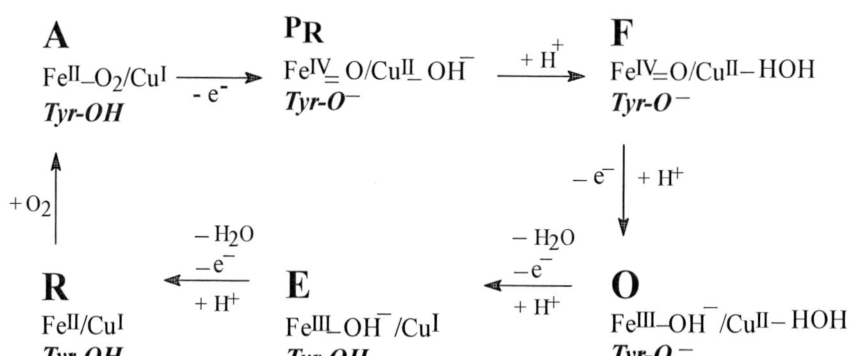

Modified from Ref [83].

Figure 2. The protonation state of the cross-linked Tyr^{280} and the kinetically resolved intermediates in the catalytic cycle of C*c*O.

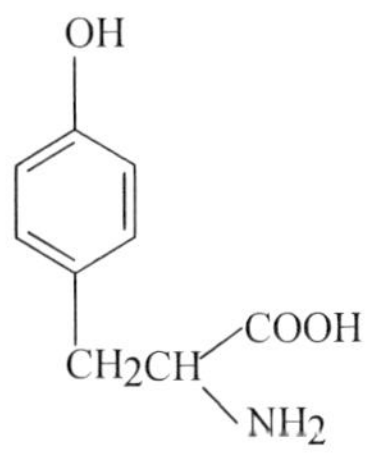

Tyr(Y)

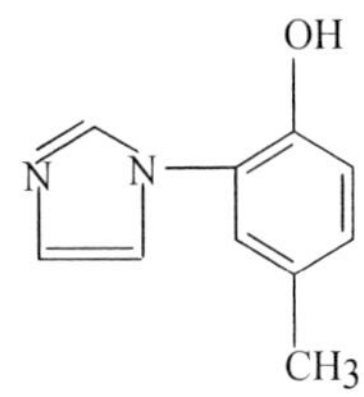

2'- (1- imidazolyl) - 4 - methylphenol

(YH)

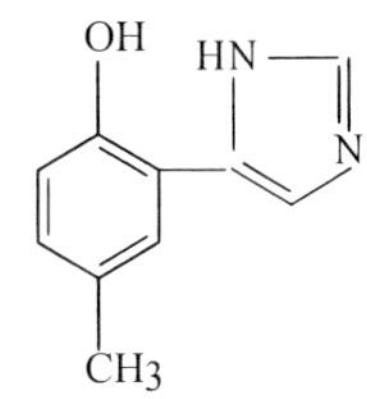

2 - (5 - imidazolyl) - 4 - methylphenol

(C I)

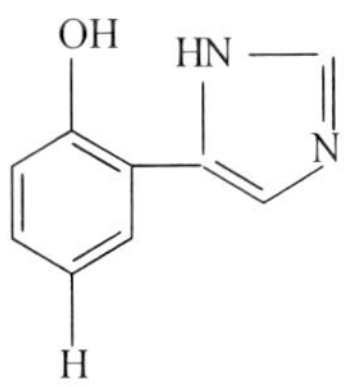

2 - (5 - imidazolyl) - 4 - H - phenol

(C II)

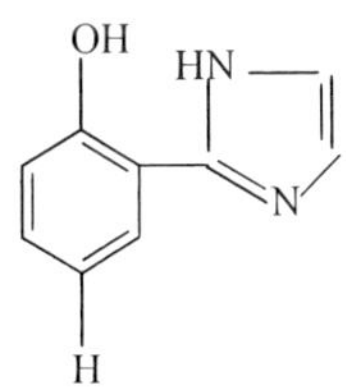

2 - (5 - imidazolyl) - 4 - H - phenol

(C III)

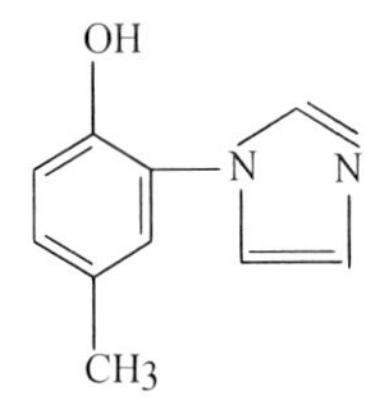

2- (1- imidazolyl) - 4 - methylphe

(C IV)

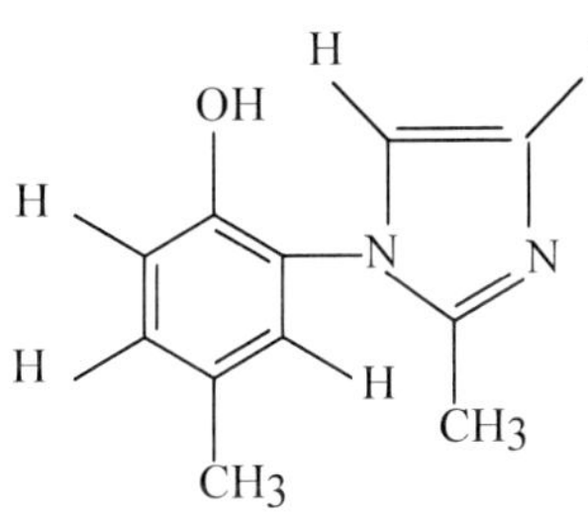

2-(2-methyl-1-imidazolyl)- 4-metylphenol

(C V)

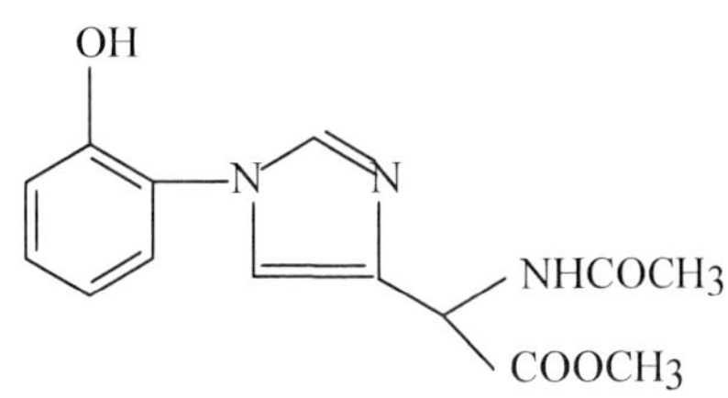

1-*o*-phenol(acetyl)histidine methyl este

(C VI)

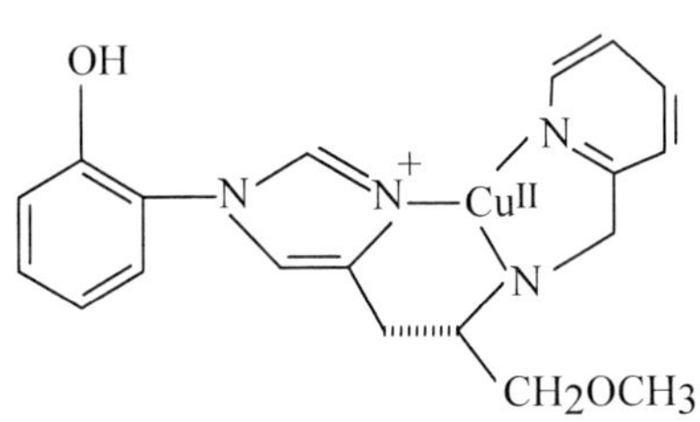

(C VII)

Figure 3. The molecular structure of Tyr and Tyr - His model compounds [85, 92-95, 97, 98].

For the 2-imidazolyl-1-yl-4-methylphenol it was found that the covalently linked imidazole ring perturbs the acidic dissociation constant of the phenol by more than 1.5 orders of magnitude. This modulation was due to stabilization of the phenolate through an inductive effect that could be even more pronounced in the protein by coordination of the imidazole to

Cu_B. It was concluded that the cross-linked Tyr-His facilitates proton delivery to the binuclear center in the C*c*O [85, 96]. In a further study 1-*ortho*-phenol(acetyl)histidine methyl ester, a cross-linked Tyr-His compound was synthesized as a chemical analogue of the active site of C*c*O [93]. The structure of the cross-linked compound was investigated by IR, ^{1}H and ^{13}C NMR, mass spectrometry and single-crystal X-ray analysis. The shifts in the FT-IR spectra, that where associated with the oxidation of the cross-linked compound, enabled the detection of significant perturbation of the phenoxyl radical vibrational bands [93]. A model compound of Tyr^{244}-His^{240} of bovine C*c*O was synthesized and analysed by UV resonance Raman, UV absorption and pH titration [94]. Some of the imidazole vibrations were resonance enhanced upon $\pi\pi^*$ transition of phenol, although they were not observed for the corresponding equimolar mixture of imidazole and *para*-cresol, indicating the delocalization of π electrons between the imidazole and phenol rings via the covalent linkage [94].

Herein, the attention is focused on 2′-(1-imidazolyl)-4-methylphenol (C-N bonding in *ortho*′- position at the phenyl group), a model compound for a Tyr-His covalent crosslink that is the most similar to that found in the C*c*O. The major interest refers to the Tyrosyl radicals formation in electron transfer and hydrogen abstraction, that can be evidenced from the specific environment or the specific interactions of Tyr with the specific protein environment.

2.1.2. Physico-chemical Analysis on Tyrosine and Tyrosine -Histidine Models

2.1.2.1. Steady-State and Time Resolved Fluorescence Spectroscopy

Fluorescence spectroscopy is a useful technique for studies on the dynamics of conformational changes in proteins. Tyrosine is one of three naturally occurring amino acids besides Phenylalanine and Tryptophan that contributes to ultraviolet fluorescence [99]. Tyrosine emission in water occurs at 303 nm and is relatively insensitive towards solvent polarity. Tyrosine is often described as a simple fluorophore. In the case of the Tyrosine zwitterion and Tyrosine derivatives with an ionized carboxyl group, mono-exponential fluorescence decay is reported, with a fluorescence lifetime of about 3.3–3.4 ns [100, 101] and a fluorescence quantum yield equal to 0.14. It was found that after the incorporation of Tyrosine into a peptide chain, its fluorescence quantum yield decreases, whereas the fluorescence intensity decay becomes heterogeneous [99, 100]. The influence of the conversion of a carboxyl group into an amide group or of the amino group into the *N*-acetyl group of Tyrosine or of its analogues on their fluorescence spectroscopic properties was widely investigated [102, 103].

The fluorescence quenching mechanism of Tyrosine derivatives is based on the dependence of the quenching efficiency on the distance between the phenol group of the Tyrosine residue and its amide group [103]. Furthermore the dependence of the electron transfer rate constant on the ionization potential of the excited fluorophore was found for a series of amidated Tyrosine analogues and their derivatives [102]. Amino group acetylation of Tyrosine or of its analogues causes a moderately decreased fluorescence quantum yield and fluorescence lifetime, the fluorescence intensity decay, however, is still mono-exponential [100]. The methylation of the hydroxyl group on the phenol ring of the Tyrosine may increase both, the fluorescence quantum yield and the fluorescence lifetime, but it does not change the character of the fluorescence intensity decay itself [100, 103].

2.1.2.2. The Effect of pH on Tyrosine and Related Compounds

The influence of the pH on the fluorescence of Tyrosine, Tryptophan and related compounds was studied by White [104]. The protonation of the Tyrosine carboxyl group decreases the fluorescence quantum yield [105] and the fluorescence lifetime [106, 107]. In the pH range from 4 to 8, the fluorescence quantum yield of Tyrosine is constant and the decay is mono-exponential with a lifetime of about 3.4 ns. This indicates that the excited-state proton transfer from phenol hydroxyl to water, though thermodynamically allowed, is too slow to affect the fluorescence decay [100, 108]. For pH > 8, the fluorescence quantum yield decreases, while the fluorescence lifetime remains constant due to the Tyrosinate formation; Tyrosinate presents a negligible fluorescence [109, 110]. The pK_a value of the carboxyl group of the 3-(4-hydroxy)-propionic acid, an analogue of Tyrosine devoid of an amino group, is 4.5 [108] or 4.9 [111], which indicates that the presence of the protonated amino group has a strong influence on the acid–base properties of the carboxyl group. A direct interaction between the carbonyl group and the phenol ring, resulting in quenching of the excited state and lowering the pK_a of the carboxyl group of Tyrosine in the excited state was suggested by Pal et al. [106] and Fayet and Whal [110]. Based on differences between pK_a values in the ground and excited states for Tyrosine isomers (*o*-, *m* and *p*-Tyrosine), Pal et al. [106] suggested that the interaction of the carboxyl group with the chromophore via the hydroxyl group of the phenol ring exists in the excited state.

The acidity of the carboxyl group of Tyrosine and its analogues and derivatives was investigated by steady state fluorescence spectroscopy [105], revealing that the methylation of the hydroxyl group of the phenol ring, as well as the position of the carboxyl group with respect to the phenol ring (Tyrosine, β-Tyrosine), had a minor influence on the value of pK_a. The introduction of an additional hydroxyl group into a phenol ring (Dopa) slightly increased the acidity of the carboxyl group compared to Tyrosine, whereas the replacement of the α-H atom in Dopa by an α-methyl group caused an increase of the pK_a value when directly compared to the value observed for Tyrosine [105]. The photophysical properties of Tyrosine at low pH range were also investigated [112] and it was found by global analysis of the fluorescence intensity decay of Tyrosine at different pH values that the quenching of Tyrosine fluorescence by H_3O^+ ions is connected to the presence of a protonated carboxyl group.

A fluorescence study concerning Trp-His interactions in the anantin peptide and in solution [113] showed that imidazole is a better quencher in the protonated form, and that the nature of the solvent is involved even in the quenching by unprotonated imidazole.

2.1.2.3. The Influence of Substituents on the Absorption and Fluorescence Behavior of Tyrosine and Tyrosine-Histidine Model Compounds

The spectral and kinetic properties caused by *para*-substituted phenyl group as well as the influence of C–C/C–N bonding in Phenol-Imidazole linkage and of the N position in the imidazole group, at physiological pH, have been reported [97]. All four Tyr-His model compounds were found to be fluorescent, the fluorescence properties depending on the molecular structure of the model compound, the C=N conjugation as well as on the stability of the model compounds in the excited state.

Despite the relatively similar molecular structure the Tyr-His model compounds, clear differences have been reported for their experimental UV-Vis absorption spectra including variations in shape and position of the characteristic bands, 270-350 nm, Table 1 [97].

2.1.2.4. The Influence of the C-C and C-N Bondings, in Phenol-Imidazole Linkage

The replacement of the C-N with a C-C bonding in the Phenol-Imidazole linkage seems to have an influence on the position of the maximum of the absorption spectrum and on the shape of the band. In the absorption spectra of the YH model compound it was seen at 283 nm. For the C I model compound, with a C-C bonding Phenol-Imidazole, the absorption maximum displayed a red-shift (18 nm) and the ε value for the C I compound that was found to be higher than the ε value of the YH compound [97].

2.1.2.5. Para-methyl Substitution at the Phenol Ring by H-Atom

The substitution by a methyl group (YH and C I model compounds) and by a hydrogen atom (C II and C III model compounds) extends the conjugation of the π-electrons from the imidazole ring to the benzene ring. A bathochromic shift between C II and C III model compounds of $\Delta_{max} \sim 6$ nm for the low – energy band ($S_0 \rightarrow S_1^*$ transition) was seen. For the C III compound, a more intense shoulder occurred at 273 nm and another absorption band at 345 nm was observed [97].

The position of the absorbance maxima is characteristic for each Tyr-His model compound. Batochromic shifts are induced by deprotonation of the phenol hydroxyl group; a process that differently takes place as a function of the model compound. The influence of *para*-methyl substitution at the phenol ring as well as the C-N/C-C bonding, in Phenol-Imidazole linkage, reduces the extension of conjugation, and a shift of the long wavelength absorption band, from 301 nm (C I) to 283 nm (YH) was observed. In the YH model compound the shoulder at 237 nm, characteristic for the anionic state was assigned to the La ($^1B_{1u}$) transition of *para*-substituted phenol. The spectrophotometric titration of the one-electron oxidation process of the C IV model compound, was found to be affected by three different dissociation constants, Ka, with the following acid dissociation constants: $pKa^1 = 8.90 \pm 0.08$ ($ImPhO^-$); $pKa^2 = 5.7 \pm 0.20$ (ImPhOH) and $pKa^3 = 4.80 \pm 0.24$ ($ImPhO^\bullet$) [85].

2.1.2.6. The N Position in the Imidazole Group

The position of the maximum in the absorption spectra of Tyr-His model compounds in direct comparison to free Tyr, may be regarded as a marker for Tyr-His containing proteins, Table 1 [97].

In the experimental data previously reported [98] no significant changes, by means of the solvent interactions, were found in the case of YH in EtOH-water and in EtOH-HCl mixtures. However a decrease in intensity was found. The same behavior was found for YH in acetonitrile and protonated in a solution of 0.1 M HCl (absorption band at 286 nm) as well as deprotonated in a solution of 0.1 M NaOH (absorption band at 313 nm) [98].

2.1.2.7. Fluorescence Decay Analysis on Tyr-His Model Compounds

Beside pH or the buffer solution, Tyr fluorescence may be quenched by an external factor, like Imidazole. This quenching was described as a dynamic process [97]. The bimolecular extinction rate constant, k_q, of the process was calculated, considering the excited-state lifetime of the Tyr fluorescence, τ_0, being 3.25 ns. Values of $K_{SV} = 22$ M^{-1} and $k_q = 0.68 \times 10^{10}$ $M^{-1}s^{-1}$ were obtained [97], indicating that Tyr is accessible to quenching by Imidazole. For Tyr, and small Tyr-containing peptides, the values of the bimolecular extinction rate constant, k_q, are near 1.2×10^{10} $M^{-1}s^{-1}$ [114, 115].

The excited - state proton transfer from Tyr to Imidazole takes place, and because the tyrosinate was found to be weakly fluorescent only [116], the ionization effectively quenchend the emission from Tyr. Im is a better quencher in the protonated than in the unprotonated form. This is explained by the formation of a non-fluorescent charge transfer complex between indole and Imidazole [113, 117].

The fluorescence decay curves at emission wavelengths of 320 nm revealed that the fluorescence lifetime values, τ, for all studied Tyr-His model compounds are lower than the fluorescence lifetime value of free Tyr. Tyr has a fluorescence decay time, $\tau = 3.38$ ns and for the model compounds a lifetime value of 3.25 ns, has been found [108]. Fluorescence lifetime and quantum yield for Tyr incorporated into a protein strongly depend on micro surrounding and rotamers population [100, 118]. For the C III compound, the τ value was seen to be smaller (1.58 ns) in comparison with the τ value of C II (2.48 ns), due to the extension of the π-electron conjugation. For the C I compound, the τ value was reported to be higher, 2.66 ns. This could be explained by the fact that the molecular structure of C I is better involved in the hydrogen bonds formation, thus the stability of the compound to the UV radiation during the lifetime measurement is higher. In Figure 4, the position of the H –bond in the C I model compound is presented. Due to the *para*-substituted phenol ring by a hydrogen atom (which could give more stability) the hydroxyl group of the C II and C III model compounds is less involved in forming a strong H-bond. The global photophysical properties of the Tyr-His model compounds were summarized in Table 1 [97].

Table 1. Absorption (λ_{Abs} and ε), Emission (λ_{em} and Fluorescence Quantum Yield (ϕ_f)), Stokes Shift ($\Delta\nu$) and the fluorescence lifetimes of the Tyr-His model compounds, in 20 mM Tris-HCl buffer, pH 7.4. The compound concentration is $5x10^{-5}$M in absorption and 3×10^{-5}M in Fluorescence, λ_{ex}= 275 nm [97]

Compound	λ_{Abs}, nm	$\varepsilon \times 10^3$, $l\ M^{-1}\ cm^{-1}$	λ_{em}, nm	$\Delta\nu$, cm^{-1}	ϕ_f	<τ> ns
Y	223		304	3357	0.14*	3.25
	274	1.0				
YH	237 (sh)		328	4568	0.08	2.03
	283	1.2				
C I	212		330	3012	0.12	2.66
	254					
	263 (sh)					
	301	2.6				
C II	210		330	3549	0.10	2.48
	254					
	262 (sh)					
	295	1.4				
C III	213		334	3236	0.06	1.58
	263		405			
	273 (sh)					
	301	3.6				
	345					

* Value from R.F. Chen [119].

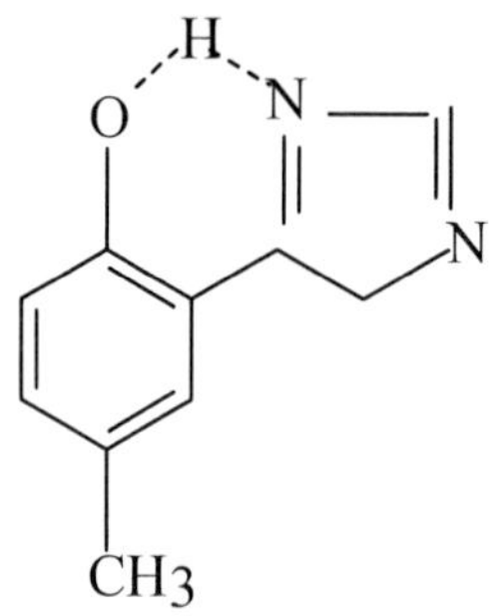

Figure 4. The position of H-bond in the chemical structure of the C I compound [97].

Guzow et *al.* [120] found that the fluorescence decay of Tyr derivatives in acetonitrile solution is monoexponential and, in contrast to the aqueous solution, the amide group does not quench the fluorescence. In conclusion, the conformations stabilized by H-bonds are involved for the heterogeneous fluorescence intensity decay of Tyr derivatives in aqueous solution [120].

Also solvent polarity was found to play an important role in the YH fluorescence quenching mechanism in the Tyr emission for the molecule where the His covalently linked to Tyr [98]. The solvent effect is partially due to the solvophobic interaction as well as to charge transfer interactions. It is known that water has a high dielectric constant ($\varepsilon = 80$), and ionic charge interactions in water are relatively weak in comparison to electrostatic interactions in the interior of a protein, where the dielectric constant ($\varepsilon = 2 - 40$) has approximately a factor of 1 : 40 directly compared to the value for water, 1 : 2 respectively. The ε value of ACN is 36. Consequently, the strength of an electrostatic interaction in the interior of a protein, where the dielectric constant is low, may be significant. The properties of the YH model compound thus corresponds to a tyrosine residue located in a buried and hydrophobic part of a membrane protein. Measurements performed in water do not correspond to the internal environment of a membrane protein. We note that electrostatic interactions in water are lower than those in other solvents because of the water's high dielectric constant, which results from the tendency of the large dipoles of water molecules to align with any electric field [121].

2.1.3. FTIR Spectroscopic Analysis

Infrared spectroscopy represents a valuable tool for structure determination of small molecules, for the investigation of protein structure, of the molecular mechanism, of protein reactions and of protein folding. [122-126]. Reaction induced infrared difference spectroscopy enables the study of the reaction induced changes even within large proteins. Valuable information on the contribution of individual amino acid side chains, cofactors, the protonation state of amino acids side chains as well as on conformational changes is available with this technique. For recent reviews see (Ref. [127] and references therein, [128, 129]).

For the Tyr-His compounds it was observed that both the imidazole as well as the phenol group are protonated/neutral in the presence of HCl and deprotonated in the presence of NaOH according to Hellwig et *al.* [96] and Wolpert et *al.* [130]. In ethanol, (Figure 5, curve a) the OH group at the Tyr ring is protonated whereas the imidazole ligand remains neutral.

Shifts and splitts in comparison to the individual amino acids have been observed, and explained with the reduced symmetry of the rings after covalent bonding.

In the the data that was reported for the deprotonated form, in Figure 5, curve c, signals at 1664 cm^{-1} and 1510-1487 cm^{-1} have been attributed to the ν(CC) vibrations of the Tyr ring [98] together with signals from the ν(CN) vibrational mode of the imidazol. The modes at 1448 cm^{-1} arise from the ν(CC) and δ_{def} (CH) vibrational modes of the Tyr, the mode at 1294 cm^{-1} from ν(C_{ring}-O) vibrational mode. The contributions at 1059/ 1074 cm^{-1}, Figure 5, curve c, may arise from ν(CN) vibrations of the imidazole ring. In direct comparison, the fully deprotonated YH compound in acetonitrile (Figure 5, curve d) was found to be highly sensitive towards the solvent and major changes in signals position and intensity have been described. In direct comparison, for the neutral form of the tyrosine, (Figure 5, curve a), shifts of all signals were evident.

The ν(CC) vibrations of the Tyr ring and the ν(CN) vibration of the imidazol ring have been suggested to be included in the signals structure at 1523 and 1495 cm^{-1} and the ν(C_{ring} – O) vibrational mode to be involved in the 1285 cm^{-1} signal. Signals below 1140 cm^{-1} arise from ν(CN) and δ NH (ring) modes. Addition of HCl protonated the imidazole ring, but did not change the tyrosine ring features.

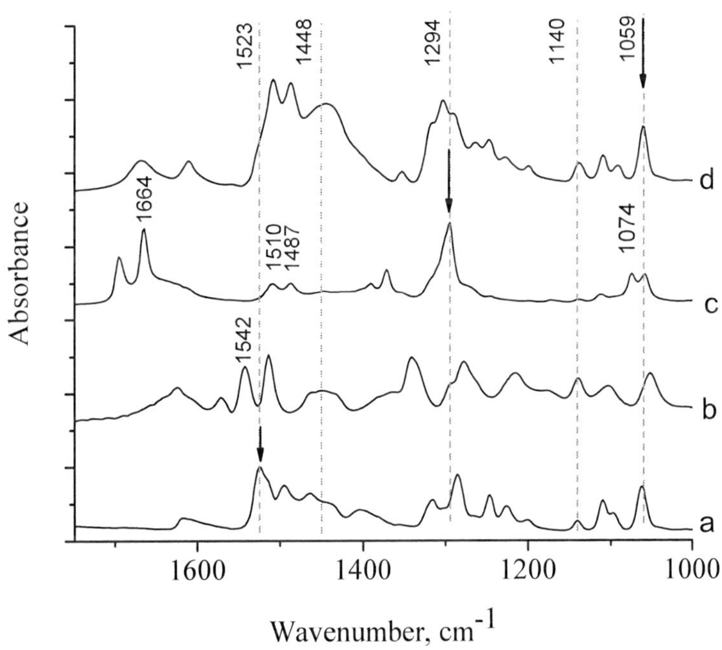

Modified from Ref [98].

Figure 5. MID-IR Absorption spectra of the YH in EtOH (a), protonated in a solution of 0.1 M HCl (b), deprotonated in a solution of 0.1 NaOH (c), YH in ACN, deprotonated in a solution of 0.1 NaOH; the YH concentration is 11 mM. The tick marks on the y-axis correspond to 0.3 absorbance units.

In Figure 5, curve b, however, a full shift of the spectrum is observed towards lower frequencies of the main vibration modes, confirming the strong coupling of the molecular vibrations. The essential features are expected in the same spectral area, 1059 cm^{-1} respectively. The ring modes of the protonated molecule may be involved in the signals at 1542 cm^{-1}.

Figure 6 shows the FTIR spectra of Tyr (A), ^{13}C(6)-tyrosine (B), His-Tyr dipeptide (C), and the His-phenol cross -linked compound, C VI model, (D). The vibrational band at 1173-1172 cm^{-1} in Tyr (Figure 6A) and in Tyr-His dipeptide (Figure 6C) was assigned to ν_{9a},(C-H) bending mode [131, 132]. The vibrational band is shifted in the C VI model compound, (Figure 6D), where one of the *o*-hydrogens has been removed and replaced by an imidazole nitrogen. Tyrosinate vibrational bands are found in the 1300-1100 cm^{-1} region, including the ν(C-O) stretching mode (ν_{7a}'), at 1260 cm^{-1} and a C-C-H bending mode (ν_{9a}), around 1170 cm^{-1} [93].

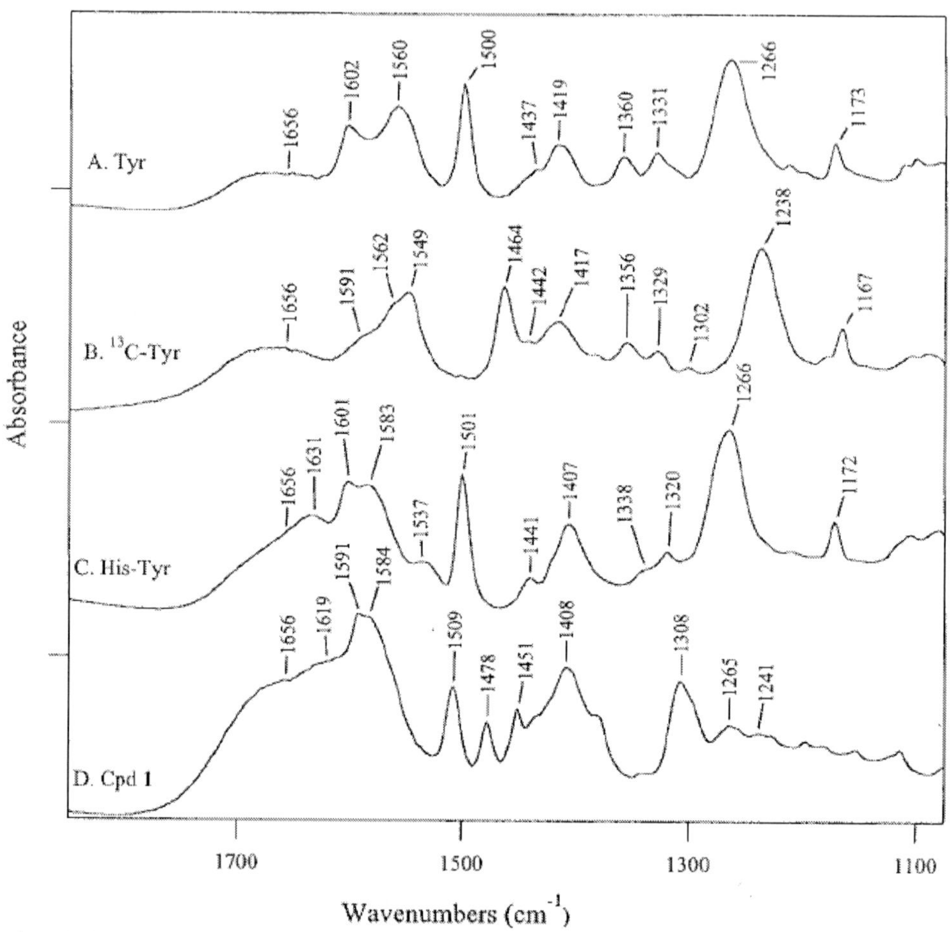

Figure 6. FTIR spectra of tyrosine (A), ^{13}C(6) -tyrosine (B), His-Tyr dipeptide (C), and His-phenol cross -linked compound (Cpd 1=C VI) (D). Samples concentration was 100 mM and in 10 mM borate-NaOH, pH 11. The borate contributions have been subtracted from these data. The tick marks on the y-axis correspond to 0.4 absorbance units.

Table 2. The main vibrational Tyrosine modes of the Tyr-His model compounds

FTIR, cm^{-1}	Assignments	References
1664/1510/1487	Tyr ring, ν(CC)	[98]
1591/1562/1560/1549/1509/1500/1464	Tyrosinate, ring ν_{8a}/ν_{19a}	[93, 131, 132]
1540-1542	Tyr-OH	[80, 87, 96, 98]
1523	Tyr ring, ν(CC)	[98]
1448	Tyr, ν(CC), δ_{def} (CH)	[98]
1437/1419/1360/1331	Tyrosinate, δ(CC)/δ(CH), ν_{19b}	[93]
1294/1285	Tyr, $\nu(C_{ring}\text{-}O)$	[98]
1269-1273	Tyr-O^-, ν(CO), ν(CC)	[127]
1260	Tyrosinate, ν (CO)	[131, 132]
1235-1270	Tyr-OH, ν(CO), ν(CC)	[127]
1173-1172	Tyrosinate δ (CH) (ν_{9a})	[131, 132]
1170	Tyrosinate, δ (CCH) (ν_{9a})	[131, 132]

In the tyrosinate spectrum (Figure 6A), the vibrational band observed at 1266 cm^{-1}, is shifted upon ^{13}C labeling (Figure 6B). A similar vibrational mode was observed in the Tyr - His dipeptide. In direct comparison, in this vibrational region, for the C VI model compound, the vibrational mode at 1265 cm^{-1} was less intense and the additional vibrational mode at 1241 cm^{-1} was observed (Figure 6D) [93]. This was in fact corresponding to unique intermolecular hydrogen-bonding interactions in the model compound, C VI [131, 132].

Vibrational modes at 1602-1601 and 1501-1500 cm^{-1} were described in both molecules, tyrosine (Figure 6A) and the Tyr-His dipeptide (Figure 6C). These spectral features downshifted to 1562 or 1549 cm^{-1} and 1464 cm^{-1}, respectively, upon $^{13}C(6)$ Tyrosine labeling (Figure 6B) [93]. The Tyr labeling experiments supported the assignment to the ν_{8a} and ν_{19a} ring stretching modes, of the Tyrosinate species. [93, 131, 132]. For the model compound C VI, the vibrational modes at 1591 and 1509 cm^{-1} have been discussed for the ν_{8a} and ν_{19a} vibrational mode (Figure 6D). It was suggested that the effect of *o*-N-substitution on the phenolate vibrational modes is a downshift of the ring mode and an upshift of the stretching vibration C-C/C-H, ν_{19} [93]. The ν_{8b} vibrational mode at 1580 cm^{-1} is shifted to 1560 cm^{-1} in the dipeptide [93]. The C=O vibration of the dipeptide and the cross-link (Figures 6 C and D respectively) are seen at 1631 cm^{-1} and 1619 cm^{-1} in Figure 6D. The CN/NH vibration of the C VI model compound, may overlap with the putative Tyrosinate ring mode around 1580 cm^{-1} (Figure 6D) [93].

Summary of data obtained by FTIR analysis of the main vibrational Tyrosine modes of the Tyr-His model compounds, is presented in Table 2.

2.1.4. Electrochemical Induced FTIR Analysis

2.1.4.1. IR modes of Tyrosyl Radical in the YH Model Compound

Electrochemically induced FTIR difference spectra have been used to identify the interactions and reorganizations of Tyr residues upon electron transfer and coupled proton transfer reactions [96] and importantly to obtain data of the radical form.

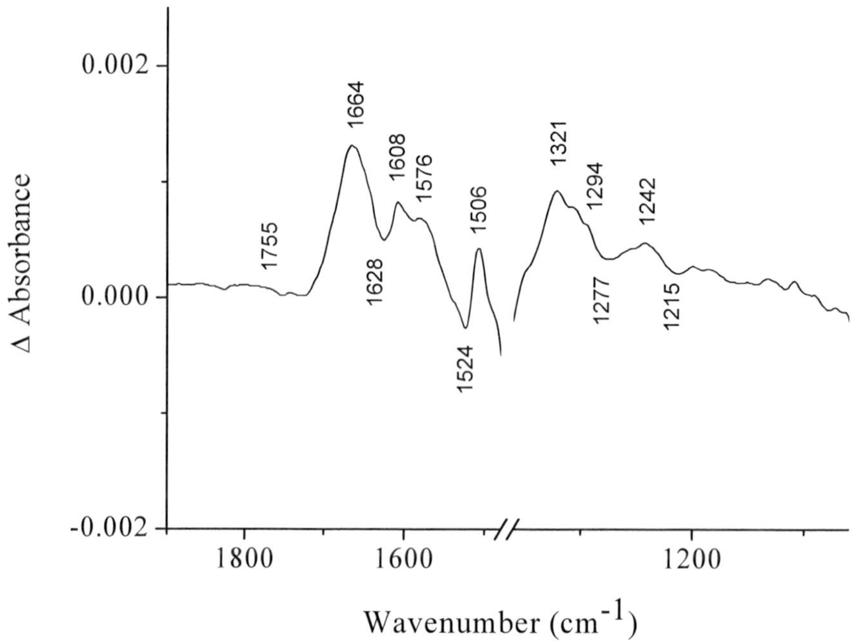

Modified from Ref [98].

Figure 7. Reduced-minus-Oxidized FTIR difference spectrum of the YH model compound in ACN for a potential step from -1.3 to +1 V (-0.63 to 1.67 V vs NHE) in the spectral range from 1800 to 1060 cm^{-1}; 0.1 M NBu_4PF_6 as supporting electrolyte. The signal region of the solvent was omitted for clarity.

Figure 7 shows the reduced minus oxidized FTIR difference spectrum of the YH model compound, for a potential step from -1.3 to 1 V, in acetonitrile and in the presence of NaOH and potentials have been given relative to the ferrocence redox couple. The positive signals shown in bold correspond to the reduced and the negative to the oxidized form of the discussed molecule. The positive signals would thus correlate to the data obtained by absorbance measurements in ACN and several signals correlate with the spectrum shown in Figure 5, curve c. This includes the vibrational ring mode at 1664 cm^{-1}, and the δ(C-O)/ν(C-C) vibrational modes around 1321 – 1294 cm^{-1}.

Signals corroborating with the oxidized form are seen with a broad feature at 1524 cm^{-1}. This position is close to the vibrational mode of the radical proposed for isolated tyrosines by Berthomieu et *al.* [133]. It is most likely including coordinates from the ν(C-O)/ν(C=C) vibration. A very small signal at 1755 cm^{-1} is seen. It may be due to the radical that is present in the carbonyl conformer. Signals in the lower frequency range have been described at 1277 and 1215 cm^{-1}[98].

2.1.4.2. Solvent Effect

In the oxidized minus reduced FTIR difference spectrum of the YH model compound described for a potential step from 0 to 0.7 V, in 20 mM Tris –HCl pH 7.4, the oxidized ν(C=O) stretching mode Tyr was seen at 1717 cm^{-1}, whereas the contribution appeared at 1489/1519 cm^{-1} were assigned to the Tyrosyl radical formation, Figure 8, [97]. This mode was also noticed in case of redox induced data obtained for free Tyr. Another prominent mode was described at 1598 cm^{-1} which is specific to the ν(C-C) Tyr ring [96]. The contribution at 1489/1519 cm^{-1} being assigned to the Tyrosyl radical formation [97].

Berthomieu et *al.* [133] observed that the ν(C-O) mode at 1480-1515 cm^{-1} is a sensitive marker of the hydrogen bonding status of substituted- phenoxyl and Tyrosyl radicals by Resonance Raman and FTIR spectroscopies.

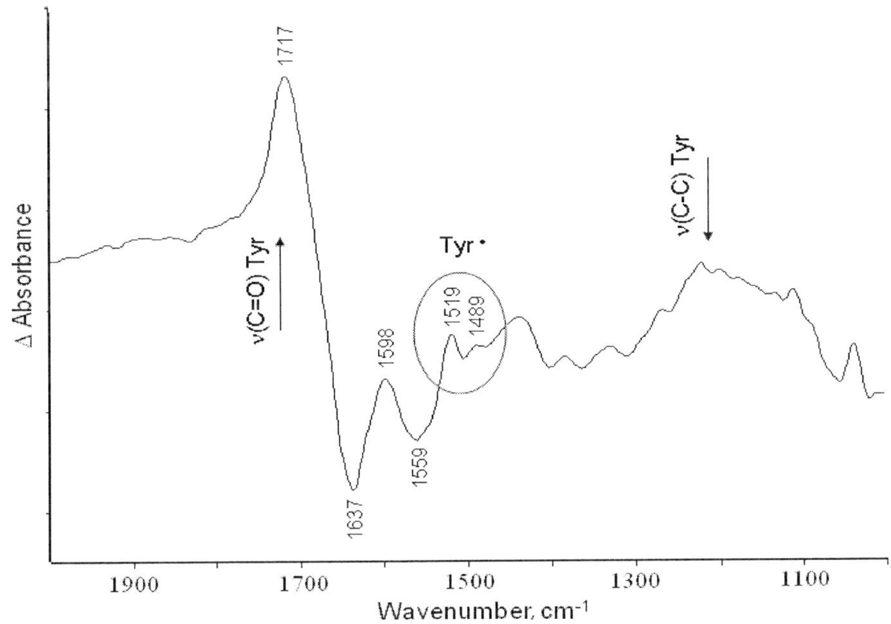

Modified from Ref [97].

Figure 8. Oxidized minus Reduced FTIR difference spectrum of the YH model compound for the potential step from 0 to 0.7 V (vs. Ag/AgCl/3 M KCl) at pH 7.4.

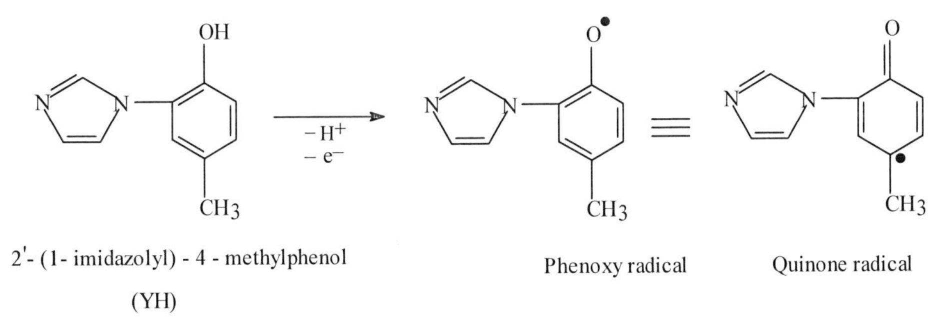

Modified from Ref [98].

Figure 9. Tyrosyl radical formation.

The possible electron- and the coupled proton transfer paths via Tyrosyl radical formation in the YH model compound is shown in Figure 9. The presence of the carbonyl function was essentially excluded in experiments performed in acetonitrile however, it was found in aqueous solution. The data gave an insight in the IR spectra of both conformers.

FTIR spectral analysis on the Tyr-His crosslink model compounds in direct comparison to those on the cytochrome *c* oxidase demonstrated the presence of a protonated form of the cross-linked tyrosine.

3. Conclusion

Tyrosine is one of the most important amino acid residues in protein structures. In this review, we discuss this amino acid together with posttranslationally modified variants and their role in the catalytic cycle. The results from different techniques are overviewed including fluorescence and infrared spectroscopies and the study on a number of model compounds Specific signals have been described for the covalent linkage and it was possible to conclude on the role of this linkage for electron transfer mechanism.

Acknowledgments

Dr. Youssef El Khoury, Institute de Chimie, Strasbourg, for the preparation of Figure 1.

References

[1] D. Voet, J.G. Voet, C.W. Prat, Fundamentals of Biochemistry, John Wiley and Sons Inc., 1998.

[2] T.E. Cerighton, Proteins, second ed., W.H. Freeman and Company, New York, 1993.

[3] D.E. Metzler, Biochemistry, second ed., vol. 1, Harcourt Academic Press, 2001.

[4] J.T. Pelton, L.R. McLean (2000), Spectroscopic methods for analysis of protein secondary structure, *Anal. Biochem.*, 277, 167-176.

[5] F.A. Armstrong, In *Bioelectrochem;* Ed. Wilson, G. S. Wiley-VCH: Weinheim, 2002; pp 11–29.

[6] L. Havran, S. Billova, E. Palecek (2004), Electroactivity of avidin and streptavidin. Avidin signals at mercury and carbon electrodes respond to biotin binding. *Electroanalysis,* 16, 1139–1148.

[7] E. Palecek, V. Ostatna (2007), Electroactivity of nonconjugated proteins and peptides. Towards electroanalysis of all proteins. *Electroanalysis*, 19, 2383-2403.

[8] M. Tomschik, L. Havran, M. Fojta, E. Palecek (1998), Constant current chronopotentiometric stripping analysis of bioactive peptides at mercury and carbon Electrodes. *Electroanalysis,* 10, 403-409.

[9] R. Brdicka (1933), *Collect. Czech. Chem. Commun.,* 5, 112–127.

[10] S. Deller, P. Macheroux, S. Sollner (2008), Flavin-dependent quinone reductases, *Cell. Mol. Life Sci.*, 65, 141-160.

[11] S.M Janes, D. Mu, D. Wemmer, A. J. Smith, S. Kaur, D. Maltby, A.L. Burlingame, J.P. Klinman (1990), A new redox cofactor in eukaryotic enzymes; 6-hydroxydopa at the active site of bovine serum amine oxidase, *Science*, 248, 981-987.

[12] N. M. Okeley, W. A. van der Donk (2001), Corrigendum to: 'Novel cofactors via post-translational modifications, *Chemistry & Biology*, 8 (1), 97.
[13] J.A Reynaud, B. Malfoy, A. Bere (1980), The electrochemical oxidation of three proteins: RNAase A, bovine serum albumin and concanavalin A at solid electrodes. *J. Electroanal. Chem.,* 116, 595–606.
[14] V. Brabec, I. Schindlerova (1981), Electrochemical behaviour of proteins at graphite electrodes Part III. The effect of protein adsorption. *Bioelectrochem. Bioenerg.,* 8, 451–458.
[15] V. Brabec, V. Mornstein (1980), Electrochemical behaviour of proteins at graphite electrodes II. Electrooxidation of amino acids. *Biophys. Chem.,* 12, 159–165.
[16] N.C. Reynolds, B.M. Kissela, L.H. Fleming (1995), The voltammetry of neuropeptides containing L-tyrosine. *Electroanalysis,* 7, 1177–1181.
[17] X. Cai, G. Rivas, P.A.M. Farias, H. Shiraishi, J. Wang, E. Palecek (1996), Potentiometric stripping analysis of biologically important bioactive peptides at carbon electrodes. *Anal. Chim. Acta.* 332, 49–57.
[18] G. Herzog, D.W.M. Arrigan (2007), Electrochemical strategies for label-free detection of amino acids, peptides and proteins. *Analyst,* 132, 615-632.
[19] R.C. Sealy, L. Harman, P.R. West, R.P. Mason (1985), The electron spin resonance spectrum of the tyrosyl radical., *J. Am. Chem. Soc.* 107, 3401–3406.
[20] S.S. Lehrer, G.D. Fasman (1967), Ultraviolet irradiation effects in poly-L-tyrosine and model compounds. Identification of bityrosine as a photoproduct. Biochemistry 6, 757–767.
[21] S. O. Anderson (1996), Covalent cross-links in a structural protein, resilin. *Acta Physiol. Scand.* 66, 1–81.
[22] H.I. Joschek, S.I. Miller (1966), Photooxidation of phenol, cresols, and dihydroxybenzenes. *J. Am. Chem. Soc.* 88, 3273–3281
[23] H. Musso (1963), Oxidations of phenols. *Angew. Chem. Int. Ed. Engl.* 2, 723–735.
[24] A.J. Gross, I.W. Sizer (1959), The oxidation of tyramine, tyrosine, and related compounds by peroxidase, *J. Biol. Chem.* 234, 1611–1614.
[25] T. Deits, M. Farrance, E.S. Kay, L. Medill, E.E. Turner, P.J. Weidman, B.M. Shapiro, (1984), Purification and properties of ovoperoxidase, the enzyme responsible for hardening the fertilization membrane of the sea urchin egg, *J. Biol. Chem.* 259, 13525–13533.
[26] C.A. Foerder, B.M. Shapiro (1977), Release of ovoperoxidase from sea urchin eggs hardens the fertilization membrane with tyrosine crosslinks, *Proc. Natl. Acad. Sci. USA,* 74, 4214–4218.
[27] P. Briza, G. Winkler, H. Kalchhauser, M. Breitenbach (1986), Dityrosine is a prominent component of the yeast ascospore wall. A proof of its structure, *J. Biol. Chem.* 261, 4288–4294.
[28] P. Briza, M. Breitenbach, A. Ellinger, J. Segall (1990), Isolation of two developmentally regulated genes involved in spore wall maturation in *Saccharomyces cerevisiae*. *Genes Dev.* 4, 1775–1789.
[29] S.K. Hanks, A.M. Quinn, T. Hunter (1988), The protein kinase family: conserved features and deduced phylogeny of the catalytic domains, *Science* 241, 42-52.
[30] J. Herrington, C. Carter-Su (2001), Signaling pathways activated by the growth hormone receptor, *Trends Endocrinol. Metab.*, 12 (6), 252-257.

[31] K. Uvdal, J. Ekeroth, P. Konradsson, B.J. Liedberg (2003), Tyrosine derivatives assembled on gold, *Coll. Interf. Sci.*, 260 (2), 361-366.

[32] D.R. Kaplan, F.D. Miller (2000), Neurotrophyn signal transduction in the nervous system, *Curr. Opin. Neurobiol.*, 10, 381-391.

[33] J. Stubbe, W.A. van der Donk (1998), Protein Radicals in Enzyme Catalysis, *Chem. Rev.*, 98, 705– 762.

[34] G. Backes, M. Sahlin, B.-M. Sjöberg, T.M. Loehr, J. Sanders-Loehr (1989), Resonance Raman spectroscopy of ribonucleotide reductase. Evidence for a deprotonated tyrosyl radical and photochemistry of the binuclear iron center, *Biochemistry*, 28, 1923– 1929.

[35] J.R. Bollinger, D.E. Edmondson, B.H. Huynh, J. Filley, J.R. Norton, J. Stubbe (1990), Mechanism of assembly of the tyrosyl radical-dinuclear iron cluster cofactor of ribonucleotide reductase, *Science*, 253, 292– 298.

[36] B.A. Barry, G.T. Babcock (1987), Tyrosine radicals are involved in the photosynthetic oxygen-evolving system, *Proc. Natl. Acad. Sci. U. S. A.*, 84, 7099– 7103.

[37] B.A. Barry, M.K. El-Deeb, P.O. Sandusky, G.T. Babcock (1990), Tyrosine radicals in photosystem II and related model compounds. Characterization by isotopic labeling and EPR spectroscopy, *J. Biol. Chem.*, 265, 20139– 20143.

[38] R. Karthein, R. Dietz, W. Nastainczyk, H.H. Ruf (1988), Higher oxidation states of prostaglandin H synthase. EPR study of a transient tyrosyl radical in the enzyme during the peroxidase reaction, *Eur. J. Biochem.*, 171, 313– 320.

[39] A.L. Tsai, L.C. His, R.J. Kulmacz, G. Palmer, W.L. Smith (1994), Characterization of the tyrosyl radicals in bovine prostaglandin H synthase-1 by isotope replacement and site-directed mutagenesis, *J. Biol. Chem.*, 269, 5085– 5091.

[40] A. Tsai, R.J. Kulmacz, G. Palmer (1995), Spectroscopic evidence for reaction of prostaglandin H synthase-1 tyrosyl radical with arachidonic acid, *J. Biol. Chem.*, 270, 10503–10508.

[41] A. Tsai, G. Wu, G. Palmer, B. Bambai, J.A. Koehni, P.J. Marshalli, R.J. Kulmacz (1999), Rapid kinetics of tyrosyl radical formation and heme redox state changes in Prostaglandin H synthase-1 and -2, *J. Biol. Chem.* 274, 21695–21700.

[42] M.G. Malkowski, S.L. Ginell, W.L. Smith, R.M. Garavito (2000), The productive conformation of arachidonic acid bound to prostaglandin synthase, *Science,* 289, 1933–1937.

[43] B.H. Dunford, *Heme Peroxidases*, Wiley, New York, 1999.

[44] Aubert, P. Mathis, A.P. Eker, K. Brettel (1999), Intraprotein electron transfer between tyrosine and tryptophan in DNA photolyase from *Anacystis nidulans*, *Proc. Natl. Acad. Sci. U. S. A.,* 96, 5423– 5427.

[45] Giovani, M. Byrdin, M. Ahmad, K. Brettel (2004), Light-induced electron transfer in a cryptochrome blue-light photoreceptor, *Nat. Struct. Biol.*, 10, 489– 490.

[46] M.M. Whittaker, J.M. Whittaker (1990), A tyrosine-derived free radical in apogalactose oxidase, *J. Biol. Chem.*, 265, 9610–9613.

[47] G.T. Babcock, M.K. El-Deeb, P.O. Sandusky, M.M. Whittaker, J.W. Whittaker (1992), Electron paramagnetic resonance and electron nuclear double resonance spectroscopies of the radical site in galactose oxidase and of thioether-substituted phenol model compounds, *J. Am. Chem. Soc.,* 114, 3727– 3734.

[48] M.M. Whittaker, Y.Y. Chuang, J.W. Whittaker (1993), Models for the redox active site in galactose oxidase, *J. Am. Chem. Soc.,* 115, 10029– 10035.

[49] M.M. Whittaker, P.J. Kersten, N. Nakamura, J. Sanders-Loehr, E.S. Schweizer, J.W. Whittaker (1996), Glyoxal oxidase from *Phanerochaete chrysosporium* is a new radical-copper oxidase, *J. Biol. Chem.*, 271, 681– 687.

[50] N. Ito, S.E.V. Phillips, C. Stevens, Z.B. Ogel, M.J. McPherson, J.N. Keen, K.D.S. Yadav, P.F. Knowles (1991), Novel thioether bond revealed by a 1.7 A crystal structure of galactose oxidase, *Nature,* 350, 87– 90.

[51] J.W. Whittaker (2003), Free radical catalysis by galactose oxidase, *Chem. Rev.*, 103, 2347– 2363.

[52] M.M. Whittaker, J. Whittaker (2001), Catalytic Reaction profile for alcohol oxidation by galactose oxidase, *Biochemistry,* 40, 7140–7148.

[53] G.T. Babcock, The oxygen-evolving complex in photosystem II as a metallo-radical enzyme, in: P. Mathis (Ed.), *Photosynthesis: From Light to Biosphere*, vol II. Kluwer Acad. Pub., Dordrecht, pp. 209– 215, 1995.

[54] Tommos, G.T. Babcock (1998), Oxygen production in nature: a light-driven metalloradical enzyme process, *Acc. Chem. Res.*, 31, 18–25.

[55] J.S. Vrettos, J. Limburg, G.W. Brudvig (2001), Mechanism of photosynthetic water oxidation: combining biophysical studies of photosystem II with inorganic model chemistry, *Biochim. Biophys. Acta,* 1503, 229– 245.

[56] R. Matsuzaki, T. Fukui, H. Sato, Y. Ozaki, K. Tanizawa (1994), Generation of the topa quinone cofactor in bacterial monoamine oxidase by cupric ion-dependent autooxidation of a specific tyrosyl residue, *FEBS Lett.*, 351, 360– 364.

[57] Cai, J.P. Klinman (1994), Copper amine oxidase: heterologous expression, purification, and characterization of an active enzyme in *Saccharomyces cerevisiae, Biochemistry,* 33, 337647 - 337653.

[58] C. Ostermeier, A. Harrenga, U. Ermler, H. Michel (1997), Structure at 2.7 A resolution of the Paracoccus denitrificans two-subunit cytochrome *c* oxidase complexed with an antibody FV fragment, *Proc. Natl. Acad. Sci. U. S. A.*, 94, 10547– 10553.

[59] S. Yoshikawa, K. Shinzawa-Itoh, R. Nakashima, R. Yaono, E. Yamashita, N. Inoue, M. Yao, M.J. Fei, C.P. Libeu, T. Mizushima, H. Yamaguchi, T. Tomozaki, T. Tsukihara (1998), Redox-coupled crystal structural changes in bovine heart cytochrome c oxidase, *Science,* 280, 1723– 1729.

[60] S. Iwata, C. Ostermeier, B. Ludwig, H. Michel (1995), Structure at 2.8 A resolution of cytochrome *c* oxidase from *Paracoccus denitrificans*, *Nature,* 376, 660– 669.

[61] T. Tsukihara, H. Aoyama, E. Yamashita, T. Tomizaki, H. Yamaguchi, K. Shinzawa-Itoh, R. Nakashima, R. Yaono, S. Yoshikawa (1995), Structures of metal sites of oxidized bovine heart cytochrome *c* oxidase at 2.8 Å, *Science*, 269, 1069– 1074.

[62] D.A. Proshlyakov, M.A. Pressler, C. DeMaso, J.F. Leykam, D.L. DeWitt, G.T. Babcock (2000), Oxygen activation in respiration: involvement of redox-active tyrosine 244, *Science*, 290, 1588–1591.

[63] MacMillan, A. Kannt, J. Behr, T. Prisner, H. Michel (1999), Direct evidence for a tyrosine radical in the reaction of cytochrome *c* oxidase with hydrogen peroxide, *Biochemistry,* 38, 9179– 9184.

[64] T. Uchida, T. Mogi, T. Kitagawa (2000), Resonance Raman studies of oxo intermediates in the reaction of pulsed cytochrome b_0 with hydrogen peroxide, *Biochemistry,* 39, 6669– 6678.

[65] T. Tsukihara, H. Aoyama, E. Yamashita, T. Tomizaki, H. Yamaguchi, K. Shinzawa-Itoh, R. Nakashima, R. Yaono, S. Yoshikawa (1996), The whole structure of the 13-subunit oxidized cytochrome c oxidase at 2.8 Å, *Science*, 272, 1136-1144.

[66] M. K. F Wikstrom (1977), Proton pump coupled to cytochrome *c* oxidase in mitochondria, *Nature* 266,271–273.

[67] S. Ferguson-Miller, G.T. Babcock (1996), Heme/copper terminal oxidases, *Chem Rev,* 96, 2889–2908.

[68] M. Wikstrom (2004), Cytochrome *c* oxidase: 25 years of the elusive proton pump, *Biochim. Biophys. Acta,* 1655, 241–247.

[69] R. B. Gennis (1998), Multiple proton – conducting pathways in cytochrome oxidase and a proposed role for the active site - tyrosine, *Biochim. Biophys. Acta*, 1365, 241-248.

[70] H. Michel (1998), The mechanism of proton pumping by cytochrome *c* oxidase *Proc. Natl. Acad. Sci. U.S.A., 95*, 12819-12824.

[71] H. Michel (1999), Cytochrome *c* oxidase: catalytic cycle and mechanisms of proton pumping-A reaction mechanism, *Biochemistry*, 38, 15129-15140.

[72] M. I. Verkhovsky, A. Jasaitis, M. L. Verkhovskaya, J. E. Morgan, M. Wikstrom (1999), Proton translocation by cytochrome *c* oxidase, *Nature,* 400, 480–483.

[73] T. Soulimane, G. Buse, G. P. Bourenkov, H. D. Bartunik, R. Huber, M. E. Than (2000), Structure and mechanism of the aberrant ba_3-cytochrome *c* oxidase from *Thermus thermophilus*, *EMBO J,* 19 (8), 1766-1776.

[74] M. Ek-Svensson, J. Abramson, G. Larsson, S. Tornroth, P. Brzezinski, S. Iwata (2002), The X-ray Crystal Structures of Wild-type and EQ(I-286) Mutant Cytochrome *c* Oxidases from *Rhodobacter sphaeroides*, *J. Mol. Biol.,* 321, 329-339.

[75] G. Buse, T. Soulimane, M. Dewor, H. E. Meyer, M. Blüggel (1999), Evidence for a copper coordinated histidine-tyrosine crosslink in the active site of Cytochrome oxidase, *Protein Sci., 8*, 985-990.

[76] M.S. Rogers, D.M. Dooley (2003), Cooper tyrosyl radical enzymes, *Current Opinion in Chemical Biology*, 7, 189-196.

[77] S. Iwata (1998), Structure and function of bacterial cytochrome *c* oxidase, *J. Biochemistry*, 123, 369-375.

[78] A. Smirnova, D. Zaslavsky, J. A. Fee, R. B. Gennis, P. Brzezinski (2008), Electron and proton transfer in the ba_3 oxidase from *Thermus thermophylus*, *J. Bioenerg. Biomembr.*, 40, 281-287.

[79] D. A. Proshlyakov, M. A. Pressler, G. T. Babcock (1998), Dioxygen activation and bond cleavage by mixed-valence Cytochrome *c* Oxidase, *Proc. Natl. Acad. Sci. U.S.A.,* 95, 8020-8025.

[80] R. M. Nyquist, D. Heitbrink, C. Bolwien, R. B. Gennis, J. Heberle (2003), Direct observation of protonation reactions, *Proc Natl Acad Sci U S A.,*100, 8715-8720.

[81] M. Iwaki, A. Puustinen, M. Wikström, P. R. Rich (2004), ATR-FTIR *Biochemistry,* 43, 14370-14378.

[82] M. Iwaki, A. Puustinen, M. Wikström, P. R. Rich (2006), Structural and chemical *Biochemistry,* 45, 10873-10885.

[83] E. A. Gorbikova, M. Wikstrom, M. I. Verkhovsky (2008), The protonation state of the cross-linked tyrosine during the catalytic cycle of Cytochrome *c* Oxidase, *J. Biol. Chem.*, 283, 34907-34912.

[84] E. A. Gorbikova, I. Belevich, M. Wikstrom, M. I. Verkhovsky (2008), The proton donor for O-O bond scission by cytochrome *c* oxidase, *PNAS*, 105, 10733-10737.

[85] M. McCauley, J. M. Vrtis, J. Dupont, W. A. van der Donk (2000), Insights into the functional role of the Tyrosine – Histidine linkage in Cytochrome *c* Oxidase, *J. Am. Chem. Soc.*, 122, 2403-2404.

[86] E. Pinakoulaki, U. Pfitzner, B. Ludwig, C. Varotsis (2002), The role of the cross-linked His-Tyr in the functional properties of the binuclear center in cytochrome *c* oxidase, *J Biol. Chem.*, 277, 13563−13568.

[87] F.L. Tomson, J. A. Bailey, R. B. Gennis, C. J. Unkefer, Z. Li, L. A. Silks, R. A. Martinez, R. J. Donohoe, R. B. Dyer, W. H. Woodruff (2002), Direct infrared detection of the covalently ring linked His-Tyr structure in the active site of the heme-copper oxidases, *Biochemistry,* 41, 14383−14390.

[88] C. W. Hoganson, C. Tommos (2004), The function and characteristics of tyrosyl radical cofactors, *Biochim. Biophys. Acta,* 1655, 116-122.

[89] Y. Nagano, J. G. Liu, Y. Naruta, T. Ikoma, S. Tero-Kubota, T. Kitagawa (2006), Characterization of the phenoxyl radical in model complexes for the Cu(B) site of cytochrome *J. Am. Chem. Soc.*, 128, 14560-14570.

[90] Y. Nagano, J.-G. Liu, Y. Naruta, T. Kitagawa (2005), UV resonance Raman study of model complexes of the Cu_B site of cytochrome *c* oxidase, *J. Mol. Struct.*, 735–736, 279–291.

[91] Y. Bu, R. I. Cukier (2005), Structural character and energetics of tyrosyl radical formation by electron/proton transfers of a covalently linked histidine-tyrosine: a model for cytochrome *c* oxidase, *J Phys. Chem. B,* 109, 22013-22026.

[92] D. A. Pratt, R.P. Pesavento, W. A. van der Donk (2005), Model studies of the histidine-tyrosine cross link in Cytochrome *c* Oxidase reveal the flexible substituent effect of the imidazole moiety, Organic Letters, 7, 2735-2738.

[93] A. Cappuccio, I. Ayala, G. I. Elliott, I. Szundi, J. Lewis, J. P. Konopelski, B. A. Barry, O. Einarsdottir (2002), Modeling the active site of Cytochrome Oxidase: synthesis and characterization of a cross-linked Histidine-Phenol, *J. Am. Chem. Soc.*, 124, 1750-1760.

[94] Aki, T. Ogura, Y. Naruta, T. H. Le, T. Sato, T. Kitagawa (2002), UV Resonance Raman characterization of model compounds of Tyr^{244} of bovine Cytochrome *c* Oxidase in its neutral, deprotonated anionic, and deprotonated neutral radical forms: effects of covalent binding between Tyrosine and Histidine, *J. Phys. Chem. A.*, 106, 3436 - 3444.

[95] K. N. White, I. Sen, I. Szundi, Y.R. Landaverry, L. E. Bria, J. P. Konopelski, M. M. Olmstead, O. Einarsdottir (2007), Synthesis and structural characterization of cross-linked histidine-phenol Cu(II) complexex as cytochrome *c* oxidase active site models, *Chem. Commun.*, 3252-3254.

[96] P. Hellwig, U. Pfitzner, J. Behr, B. Rost, R.P. Pesavento, W. v Donk, R.B. Gennis, H. Michel, B. Ludwig, W. Mantele (2002), Vibrational modes of tyrosines in Cytochrome *c* Oxidase from *Paracoccus denitrificants*: FTIR and Electrochemical studies on Tyr-D_4-labeled and on Tyr280His and Tyr35Phe Mutant enzymes, *Biochemistry*, 41, 9116-9125.

[97] Voicescu, M. Heinrich, P. Hellwig (2009), Steady – State and Time Resolved Fluorescence Analysis of Tyrosine-Histidine Model Compounds, *J of Fluorescence*, 19, 257-266.

[98] Voicescu, Y. El Khoury, D. Martel, M. Heinrich, P. Hellwig (2009), Spectroscopic Analysis of Tyrosine Derivatives: On the Role of the Tyrosine - Histidine Covalent Linkage in Cytochrome *c* Oxidase, *J. Phys. Chem. B*, 113, 13429-13436.

[99] J. R. Lakowicz, Principles of Fluorescence Spectroscopy, Kluwer Academic Publishers, New York, 1999.

[100] J.B.A. Ross, W.R. Laws, K.W. Roussland, H.R. Wyssbrod, Tyrosine fluorescence and phosphorescence from proteins and polypeptides, in: J.R. Lakowicz (Ed.), *Topics in Fluorescence Spectroscopy*, Vol. 3, Biochemical Applications, Plenum Press, New York, pp. 1–63, 1992.

[101] C. Seidel, A. Orth, K.-O. Greulich (1993), Electronic effects on the fluorescence of tyrosine in small peptides, *Photochem. Photobiol.* 58, 178-184.

[102] W. Wiczk, A. Rzeska, J. Łukomska, K. Stachowiak, J. Karolczak, J. Malicka, L. Łankiewicz (2001), Mechanism of fluorescence quenching of tyrosine derivatives by amide group, *Chem. Phys. Lett.*, 341, 99 -106.

[103] J. Łukomska, A. Rzeska, J. Malicka, W. Wiczk (2001), Influence of substituent on amide nitrogen atom on fluorescence efficiency quenching of Tyr (Me) by amide groups, *J. Photochem. Photobiol. A: Chem.*, 142, 135-123.

[104] A. White (1959) Effect of pH on fluorescence of Tyrosine, Tryptophan and related Compounds, *Biochem. J*, 71, 217-220.

[105] M. Szabelski, K. Guzow, A. Rzeska, J. Malika, M. Przyborowska, W. Wiczk (2002), Acidity of carboxyl group of tyrosine and its analogues and derivatives studied by steady-state fluorescence spectroscopy, *J. of Photochem. Photobiol. A: Chem*, 152, 73-78.

[106] H. Pal, D.K. Palit, T. Mukherejee, P. Mittal (1990), Some aspects of steady state and time-resolved fluorescence of tyrosine and related compounds, *J. Photochem. Photobiol. A: Chem.* 52, 391-409.

[107] K. J. Wilis, A.G. Szabo (1991), Fluorescence decay kinetics of tyrosinate and tyrosine hydrogen-bonded complexes, *J. Phys. Chem.*, 95, 1585-1589.

[108] W.R. Laws, J.B.A. Ross, H.R. Wyssbrod, J.M. Beechem, L. Brand, J.C. Sutherland (1986) Time-resolved fluorescence and proton NMR studies of tyrosine and tyrosine analogs: correlation of NMR-determined rotamer populations and fluorescence kinetics, *Biochemistry*, 25, 599-607.

[109] D.M. Rayner, D.T. Kraycarski, A.G. Szabo (1978) Excited state acid-base equilibrium of tyrosine, *Can. J. Chem.*, 56, 1238-1245.

[110] M. Fayet, P. Wahl (1971) Variation avec le pH du rendement quantique et du déclin de la fluorescence de la tyrosine. The variation in the quantum yield and fluorescence decay of tyrosine as a function of pH, *Biochim. Biophys. Acta*, 221, 102-112.

[111] J. Feitelson (1964) On the Mechanism of Fluorescence Quenching. Tyrosine and Similar Compounds, *J. Phys. Chem.*, 68, 391-397.

[112] K. Guzow, M. Szabelski, A. Rzeska, J. Karolczak, H. Sulowska, W. Wiczk (2002) Photophysical properties of tyrosine at low pH range, *Chem. Phys. Lett.*, 362, 519-526.

[113] R. Vos, Y. Engelborghs (1994) A Fluorescence study of Tryptophan-Histidine interactions in the peptide anantin and in solution, *Photochem. Photobiology*, 60, 24-32.
[114] J.R. Lakowicz, G. Weber (1973) Quenching of fluorescence by oxygen. A probe for structural fluctuations in macromolecules, *Biochemistry*,12, 4161-4170.
[115] J.R. Lakowicz, G. Weber (1973) Quenching of protein fluorescence by oxygen. Detection of structural fluctuations in proteins in the nanosecond time scale, *Biochemistry*,12, 4171-4179.
[116] J.R. Lakowicz, B.P. Maliwal (1983) Oxygen quenching and fluorescence depolarization of tyrosine residues in proteins, *J of Biological chemistry*, 258, 4794-4801.
[117] M. Shinitzky, R. Goldman (1967) Fluorometric detection of histidine-tryptophan complexes in peptides and proteins, *Eur. J. Biochem*., 3, 139-144.
[118] J.B.A. Ross, W.R. Laws, A. Buku, J.C. Sutherland, H.R. Wyssbrod (1986) Time – resolved fluorescence and ^{1}H NMR studies of tyrosyl residues oxytocin and small peptides. Correlation and of NMR determined conformations of tyrosyl residues and fluorescence decay kinetics, *Biochemistry*, 25, 607- 612.
[119] R.F. Chen (1967), Fluorescence quantum yields of tryptophan and tyrosine, *Anal. Lett.*, 1, 35-42.
[120] K. Guzow, R. Ganzynkowicz, A. Rzeska, J. Mrozek, M. Szabelski, J. Karolczak, A. Liwo, W. Wiczk (2004) Photophysical properties of tyrosine and its simple derivatives studies by time-resolved fluorescence spectroscopy, global analysis and theoretical calculations, *J. Phys. Chem. B*, 108, 3879-3889.
[121] T.E. Creighton, *Proteins: Structures and Molecular Properties*. New York, NY:W.H. Freeman and Company, 1993.
[122] J.L.R. Arrondo, A. Muga, J. Castresana, F.M. Goñi (1993), Quantitative studies of the structure of proteins in solution by Fourier-transform infrared spectroscopy, *Prog. Biophys. Mol. Biol.* 59, 23–56.
[123] F. Siebert (1995), Infrared spectroscopy applied to biochemical and biological problems, *Methods Enzymol*. 246, 501–526.
[124] E. Goormaghtigh, V. Cabiaux, J.-M. Ruysschaert (1994), Determination of soluble and membrane protein structure by Fourier transform infrared spectroscopy. I. Assignments and model compounds, *Subcell Biochem*. 23, 329–362.
[125] P.I. Haris, D. Chapman, *Analysis of polypeptide and protein structures using Fourier transform infrared spectroscopy*, in: C. Jones, B. Mulloy, A.H. Thomas (Eds.), Microscopy, Optical Spectroscopy, and Macroscopic Techniques ed., Vol. 22, Methods in Molecular Biology, Humana Press Inc, Totowa, NJ, pp. 183–202, 1994
[126] K. Gerwert, Molecular reaction mechanisms of proteins monitored by time-resolved FTIR-spectroscopy, *Biol. Chem*. 380, 931–935, 1999.
[127] A. Barth (2007), Infrared spectroscopy of proteins, *Biochim. Biophys. Acta*, 1767, 1073-1101.
[128] J. Kong, S. Yu (2007), Fourier transform infrared spectroscopic analysis of protein secondary structures, *Acta Biochimica et Biophysica Sinica,* 39(8), 549-559.
[129] R. Vogel, F. Siebert (2000), Vibrational spectroscopy as a tool for probing protein function, *Curr. Opin. Chem. Biol.* 4, 518–523.

[130] M. Wolpert, P. Helwig (2006), Infrared spectra and molecular absorption coefficients of the 20 alpha amino acids in aqueous solutions in the spectral range from 1800 to 500 cm^{-1}, *Spectrochimica Acta Part A*, 64, 987-1001.
[131] H. Takeuchi, N. Watanabe, Y. Satoh, I. Harada (1989), Effects of hydrogen bonding on the tyrosine Raman bands in the 1300-1150 cm^{-1} region, *J. Raman Spectrosc.*, 20, 233-237.
[132] A. Mukherjee, M. L. McGlashen, T.G. Spiro (1995), Ultraviolet Resonance Raman spectroscopy and general valence force field analysis of phenolate and phenoxyl radical, *J Phys. Chem*, 99 (14), 4912-4917.
[133] C. Berthomieu, R. Hienerwadel (2005) Vibrational spectroscopy to study the properties of redox-active tyrosines in photosystem II and other proteins, *Biochimica et Biophysica Acta*, 1707, 51-66.

In: Tyrosine and Aspartic Acid
Editors: J. E. Jones and D. M. Morano
ISBN: 978-1-62100-752-4

Chapter 3

Molecular Mechanism Leading to Preferential Replication of HIV-1 in CD45RO CD4+ T Lymphocytes

Brigitte Audet[1,*]**,** ***Cynthia Torresilla***[2,*]**,**
Gilles A. Robichaud[1]**,** ***Michel J. Tremblay***[1] ***and Benoit Barbeau***[2,†]
[1]Centre de Recherche en Infectiologie, Centre Hospitalier Universitaire de Québec, Pavillon CHUL, and Département de Biologie médicale, Faculté de Médecine, Université Laval, Ste-Foy, Québec, Canada
[2]Département des sciences biologiques and Centre de recherche BioMed, Université du Québec à Montréal, Montréal, Québec, Canada

Abstract

The ability of memory CD4+ T cells (CD45RO) to better replicate HIV-1 in comparison to naïve CD4+ T cells (CD45RA) have been well established. In a Jurkat cell system, we had previously suggested that the nuclear factor of activated T cells (NFAT) was a key element for the higher HIV-1 LTR activation level and the more intense HIV-1 replication observed in Jurkat cells expressing the CD45RO isoform (as compared to CD45RA-expressing cells). To further investigate on the step of the HIV-1 replicative cycle required for preferential replication in memory CD4+ T cells, primary CD4+ T lymphocytes were sorted in naïve and memory CD4+ T cell subsets using antibody-coupled magnetic beads and subsequently infected. We first confirmed that wild-type and pseudotyped luciferase-expressing HIV-1 virions showed stronger replication in memory CD4+ T lymphocytes following cellular activation. Early measurement of intracellular p24 after contact of T lymphocytes with HIV-1 virions and experiments using VSV-G-pseudotyped HIV-1 virions indicated that viral entry was comparable in both memory and naïve CD4+ T cells. Through real time PCR analysis, we demonstrated that the content in viral DNA (total or integrated) was higher in the memory CD4+ T cell

---

* Brigitte Audet and Cynthia Torresilla have contributed equally to this work.
† Address correspondence to: Benoit Barbeau (current address), Université du Québec à Montréal, Département des sciences biologiques, CP 8888, Succ. C.V., Montréal, Québec, Canada, Tel: (514) 987-3000 (ext. 4576). Fax: (514) 987-4647. E-Mail: barbeau.benoit@uqam.ca.

population upon activation. The involvement of NFAT in this process was strongly suggested by Electrophoretic Mobility Shift Assays and infection experiments conducted in the presence of the NFAT inhibitor FK506. Our results thereby argues that reverse transcription is an important step restricting HIV-1 replication to memory CD4+ T cells and that NFAT would be an important regulator of this process.

INTRODUCTION

The Human Immunodeficiency Virus type 1 (HIV-1) exhibits an important tropism for CD4+ T cells due to a high affinity interaction between the CD4 glycoprotein, which serves as the viral receptor for its gp120 envelope glycoprotein [1-3]. Following CD4 binding and interaction with its appropriate chemokine co-receptor, the gpl20 glycoprotein undergoes conformational change allowing gp41 to mediate fusion between viral and target cell membranes. Viral RNAs released in the cytoplasm are subsequently reverse transcribed and generated proviral DNAs are then incorporated into the host genomic DNA and transcribed for the production of viral proteins and eventually new virions. Steps of the viral cycle have been shown to be intimately linked to T-cell activation, most strikingly at the level of viral gene expression [4].

T cell receptor (TCR)-mediated stimulation induces tyrosine phosphorylation of a number of cellular proteins, which leads to functional activation of T cells. An important protein-tyrosine phosphatase (PTP) implicated in T cell activation is CD45 [5-7]. This transmembrane PTP is expressed in all hematopoietic cells with the exception of erythrocytes and platelets. The CD45 gene produces several isoforms and their specific expression depends on the state of differentiation of hematopoietic cells. The expression of the various CD45 isoforms arises from differences in the splicing of the exons A, B, and C. CD45RABC and CD45RO, which respectively contains or lacks all three exons, represent important markers to differentiate naïve from (central and effector) memory T cells [8, 9]. In fact, differential expression of CD45 isoforms in naïve versus memory CD4+ T cells has been important to demonstrate differences in CD4+ T cell response of both subsets to T cell activators [8, 10-12].

The importance of memory CD4+ T cells in HIV-1 pathogenesis and treatment has been greatly underscored in the recent years due to the implication of this cell subtype in maintaining a latent viral reservoir. Several experiments have shown that activated CD4+ T cells are highly susceptible to HIV-1 infection and these infected cells upon reversion to quiescent memory CD4+ T cells would thereby mainly constitute this reservoir [13, 14]. However, many studies have also shown that HIV-1 preferably replicates in memory CD4+ T cell subsets in contrast to the naïve CD4+ T cell subset [7, 15-20]. A mechanism implicating a more potent fusion of viral particles at the surface of memory CD4+ T cells has been proposed [21, 22]. Using Jurkat cell clones expressing either the CD45RABC or CD45RO isoform, we have previously reported that HIV-1 infection was more pronounced in CD45RO-expressing Jurkat cells and that this could reflect another mechanism explaining preferential replication of HIV-1 in memory CD4+ T cells [19]. Indeed, two inhibitors of calcineurin serine/threonine phosphatase activity and of NFAT activation, i.e. cyclosporin A and FK506 [23-25] profoundly diminished the difference in replication observed between both CD45 isoform-expressing Jurkat clones. Based on previous studies showing that NFAT was binding to the HIV-1 enhancer and could positively modulate its activity [26-28], we had then suggested that the mechanism of action of NFAT was transcription-dependent.

To clearly readdress the previously demonstrated HIV-1 preferential replication in CD45RO CD4+ T cells, in the present study, we have investigated the molecular mechanism underlying this phenomenon using purified naïve and memory CD4+ T lymphocytes. Our results have confirmed that higher viral replication occurs in CD45RO CD4+ T lymphocytes upon cellular activation and that this process was independent of viral entry. Our results strongly suggest that higher HIV-1 replication in memory CD4+ T cells involves a more active reverse transcription step and that NFAT would be a key element in this step of HIV-1 replication.

Experimental Procedures

Cell Culture and Purification of Memory and Naïve CD4+ T Lymphocytes

Peripheral blood mononuclear cells (PBMCs) from healthy donors were isolated by Ficoll-Hypaque gradient centrifugation. CD4+ T cells were first purified from freshly isolated PBMCs by negative selection using antibody-coated magnetic beads and the AutoMACS device (Miltenyi Biotec, Auburn, CA) according to manufacturer's instructions. Subsequently, from these isolated CD4+ T cells, a second round of negative selection were undertaken with antibody-coated magnetic beads from the CD45RA or CD45RO cell isolation kit (Miltenyi Biotec). Purified CD45RA+ and CD45RO+ T cells were cultured in RPMI 1640 supplemented with 10% Fetal Bovine Serum (FBS) (HyClone Laboratories, Logan, UT), glutamine (2 mM), penicillin G (100 units/ml), and streptomycin (100 μg/ml) and were maintained at 37°C under a 5% CO_2 humidified atmosphere. The human 293T embryonic kidney cell line was kindly provided by Dr. W.C. Greene (The J. Gladstone Institutes, San Francisco, CA) and was cultured in Dulbecco's Modified Eagle's Medium (DMEM) supplemented with 10% FBS, antibiotics and glutamine. DT30 cells are derived from the P815 mastocytoma cell line and stably express human B7.1 on their cell surface [29]. These cells also express the murine Fcγ receptor and can thus bind and cross-link soluble antibodies. This cell line was obtained from the American Type Culture Collection (ATCC) (Rockville, MD). DT30 cells were grown in RPMI 1640 supplemented with 10% FBS, antibiotics and glutamine and were fixed in 1% paraformaldehyde, washed extensively with phosphate-buffered saline (PBS), and then stored frozen at a density of 5×10^6/ml in RPMI 1640 prior to use.

Plasmids and Antibodies

The pNL4-3 proviral DNA clone is a full-length infectious HIV-1 molecular clone provided by the NIH AIDS Research and Reference Reagent Program (Rockville, MD). The pNL4-3 Luc^+Env^- vector (containing the luciferase reporter gene and deficient for envelope synthesis) and pHXB2-env (an HIV-1 envelope expression vector) were generously provided by Dr. N. R. Landau (The Salk Institute for Biological Studies, La Jolla, CA) [30]. The pHCMV-G vector expressing the broad host-range Vesicular Stomatitis Virus envelope glycoprotein G (VSV-G) has been described previously [31]. The pHXB-LUC proviral DNA vector contains the luciferase reporter gene inserted in the *nef* gene and has been provided by Dr. D. Baltimore (Rockfeller University, New York, NY) [32]. Both hybridomas producing anti-p24 (clone 31-90-25) and anti-CD3 OKT3 (specific for the ε chain of the CD3 complex) antibodies were obtained from the ATCC. Antibodies from these hybridomas were purified with mAbTrap protein G affinity columns according to the manufacturer's instructions (Amersham Biosciences Inc., Uppsala, Sweden). Antibodies directed against the CD45RABC and CD45RO isoforms (clone H100 and UCHL-1, respectively) were purchased from BD PharMingen (Mississauga, Canada). Anti-CD4 SIM.4 monoclonal antibodies were obtained from the NIH AIDS Research and Reference Reagent Program. The R-phycoerythrin (PE)-conjugated goat anti-mouse immunoglobulin G was purchased from Jackson ImmunoResearch Laboratories (West Grove, PA.).

Flow Cytometry

Expression of CD4, CD45RA and CD45RO on the surface of cells was monitored with the following antibodies: SIM.4, clone H100 and UCHL-1, respectively. Briefly, an aliquot of isolated cells (5×10^5) was washed with PBS. Pelleted cells were incubated for 30 min on ice with a saturating concentration of the primary antibody (1 μg/10^6 cells) in a final volume of 100 μl of PBS. The cells were washed twice with 500 μl of PBS and incubated for 30 min on ice with an (R)-PE-conjugated goat anti-mouse IgG (1 μg/10^6 cells) in a final volume of 100 μl of PBS. Finally, cells were washed twice with PBS and resuspended in 300 μl of PBS containing 1% (w/v) paraformaldehyde before cytofluometric analysis (EPICS Elite ESP; Coulter Electronics, Miami, FL).

Production of Virus Stocks

Virus particles were produced by calcium phosphate transfection of virus-encoding vectors in 293T cells as previously described [33, 34]. Briefly, 293T cells were plated 16 h before transfection to reach a 50-80% confluence the day of transfection and transfected with 35-40 μg of DNA using the Clontech Transfection Kit (BD Biosciences, San Jose, CA). Pseudotyped HIV-1 particles were generated by cotransfecting pNL4-3 Luc^+Env^- (25-30 μg) and envelope expression vectors (5 μg of pHCMV-G or 15 μg of pHXB2-env). Sixteen hours after transfection, the cells were washed twice with PBS and incubated for 24 h in complete DMEM culture medium. Viruses were collected at this point by filtering the culture media through a 0.45-μm pore size cellulose acetate membrane and were then concentrated by Centricon plus-20 filter devices (Millipore, Bedford, MA). Virus stocks were normalized for virion content using an in-house sensitive double antibody sandwich enzyme-linked immunosorbent assay (ELISA) specific for the major core viral p24 protein as previously described [35]. Virus stocks were aliquoted and frozen at -85°C for future use. All virus stocks underwent only one freeze-thaw cycle before their use in infection studies.

Viral Infection and Quantification

Freshly isolated CD45RA and CD45RO CD4+ T cells (4×10^6) were first inoculated with virions (0.5 pg of p24 per cell) in a volume of 1 ml of complete medium and incubated for 16-18 h at 37°C. After removal of excess viruses, cells were resuspended in fresh medium and aliquoted into 96-well microtiter plates at a concentration of 10^5 cells/200 µl/well in triplicates. Cells were subsequently stimulated with phytohemagglutinin-P (PHA-P at 3 µg/ml, Sigma, St-Louis, MO) or anti-CD3 antibodies (clone OKT3 at 3 µg/ml) added to DT30 cells (10^5 DT30/10^5 target cells). Unstimulated infected cells and uninfected cells were included as controls. For some experiments, prior to the addition of the stimuli, cells were either left untreated or were pretreated with FK506 (10 ng/ml) (Fujisawa, Osaka, Japan) for 60 min at 37 °C. Productive viral replication was assessed by measuring soluble p24 antigen released into culture supernatant. Samples of 100 µl were taken from each microculture well after 1, 3 and 5 days of stimulation, and stored at -80°C until completion of the experiment. Batch assays for p24 antigen quantification were performed by the p24 ELISA assay.

For infection experiments with luciferase-expressing HIV-1 virions, luciferase activity was monitored from day 1 to day 7 post-stimulation following a previously described protocol [36] in which 25 µl of 5X lysis buffer (25 mM Tris phosphate, pH 7.8, 2 mM dithiothreitol, 1% Triton X-100, 10% glycerol) was directly added to the cells. Luciferase activity was monitored from a 20 µl aliquot of each cell lysate with a Dynex MLX microplate luminometer (MLX Dynex Technologies, Chantilly, VA) upon addition of 100 µl of a luciferase buffer (20 mM tricine, 1.07 mM $(MgCO_3)_4 \cdot Mg(OH)_2 \cdot 5\ H_2O$, 2.67 mM $MgSO_4$, 0.1 mM EDTA, 220 µM coenzyme A, 4.7 µM D-luciferin potassium salt, 530 µM ATP, 33.3 mM dithiothreitol). Mean luciferase activity ±standard deviation (SD) was calculated using luciferase activities from three independent and similarly treated cell samples taken from the same experiment.

Detection of Intracellular P24

Isolated CD45RA and CD45RO CD4+ T lymphocytes were fixed and permeabilized according to manufacturer's instructions (Fix and Perm; An-der-Grub Bio Research, Kaumberg, Austria). Briefly, cells (5×10^5) were washed once with PBS at room temperature and cell pellets were then incubated for 15 min at room temperature with 25 µL reagent A (fixation medium) in a final volume of 125 µl of PBS.

After a single wash with 1 ml of PBS, pelleted cells were resuspended in 100 µL PBS and incubated in the presence of 25 µL reagent B (permeabilisation medium) and 1 µg/10^6 cells of anti-p24 antibody (clone 31-90-25) for 30 min on ice.

Cells were washed with 1 ml of PBS containing 0.03% Triton X-100 and incubated with a saturating concentration of (R)-PE-conjugated goat anti-mouse IgG (1 µg/10^6 cells) in a final volume of 100 µl PBS-0.03% Triton X-100 for 30 min on ice.

Cells were finally washed twice with PBS-0.03% Triton X-100 and resuspended in 300 µl of PBS containing 1% (w/v) paraformaldehyde before cytofluometric analysis.

Viral Entry Assay

Freshly isolated CD45RA and CD45RO CD4+ T cells (1 x 10^5) were incubated for 1 h at 37°C with virions (50 ng of p24) in triplicates. Cells were then trypsinized for 5 min at 37°C to remove uninternalized viruses and then washed with RPMI supplemented with 10% FBS, followed by three successive washes with PBS.

Each cell sample was lysed in 200 µL of lysis buffer (200 mM HEPES, 150 mM NaCl, 0.5% Triton) and cell extracts were evaluated for their content in HIV-1 p24 by the p24 ELISA assay.

Real-Time PCR Analysis of Total and Integrated Viral DNA

Total cellular DNA from purified CD45RA and CD45RO CD4+ T cells (5 × 10^6) harvested after 5 days of stimulation was extracted using 1 ml Trizol (Life Technologies, Grand Island, NY) according to the manufacturer's protocol. Isolated DNA was then washed twice in 0.1 M sodium citrate/10% ethanol, once in 75% ethanol and then dissolved in 8 mM NaOH. Real-time PCR was next conducted on the isolated DNA to quantify viral DNA at the level of both total and integrated viral DNA forms. For all PCR amplification conditions, the following reagents were added: ThermoPol buffer 1X (New England Biolabs Canada, Pickering, Canada), 0.4 mM dNTPs (Amersham Biosciences Corp., Piscataway, NJ), primers (1 µM) and 2 U Taq DNA polymerase (New England Biolabs Canada). Quantification of total viral DNA was performed on 50-100 ng of isolated genomic DNA by the detection of early LTR/U5 viral DNA through real-time PCR using HIV-FAM Taqman fluorogenic probe (5'-TAGTGTGTGCCCGTCTGTTGTGTGAC-3') (Biosearch Technologies Inc., Novato, CA) and the M667 (5'-GGCTAACTAGGGAACCCACTG-3') and AA55 (5'-CTGCTAGA GATTTTCCACACTGAC-3') primers. Real-time PCR was conducted in a Rotor Gene RG-3000 device (Corbett Research, Mortlake, Australia) with cycling conditions including initial denaturation (95°C for 5 min) followed by 40 cycles of denaturation (95°C for 15 s) and extension (63°C for 1 min). For quantification of viral DNA copy number, a standard curve was similarly amplified in parallel and consisted of 2343 to 6 x 10^5 pNL4-3 DNA copies diluted in 1 ng of DNA extracted from uninfected cells. For real-time PCR quantification of integrated HIV-1 DNA, the procedure reported by Suzuki *et al.* [37] was adopted, which consisted of a first round of PCR conducted in duplicate with an *Alu*-sequence-specific sense primer (5'-TCCCAGCTACTCGGGAGGCTGAGG-3') and the antisense gag-specific M661 primer (5′-CCTGCGTCGAGAGATCTCTG-3′). Amplification of 100 ng of isolated genomic DNA was conducted with the following cycling conditions: initial denaturation (94°C for 3 min) followed by 22 cycles of denaturation (94°C for 30 s), annealing (66°C for 30 s) and extension (70°C for 10 min) with a final extension of 10 min at 72°C. PCR products from uninfected and unstimulated infected cells were diluted 10-fold, while amplified products from infected and stimulated cells were diluted 100-fold and subsequently subjected to real-time PCR conditions described above. Normalization for DNA content and the efficiency of HIV-1 DNA amplification in between samples for all real-time PCR experiments was achieved though real-time PCR reactions specific for the β-actin gene using 0.4X SybrGreen I (Molecular Probes, Eugene, OR) and sense: (5'-CGTGACATTAAGGAGAAGCTGTGC-3') and antisense: (5'- CTCAGGAGGAGCAATGATCTTGAT-3') primers and the following

PCR conditions: 95°C for 5 min followed by 40 cycles of denaturation (95°C for 15 s), annealing (58°C for 20 s) and extension (72°C for 35 s). Standard curve for the real-time PCR amplification of the β-actin gene consisted of genomic DNA (set at 100 ng/μl) with a series of two-fold dilutions. The amount of HIV-1-specific DNA was calculated after normalization from duplicates independently amplified for both HIV-1 DNA and the β-actin gene and values are expressed in arbitrary units (AU).

Preparation of Nuclear Extracts and Electrophoretic Mobility Shift Assay (EMSA)

Isolated CD4+ CD45RA and CD45RO T cells (5 x 10^6) harvested after 24 h of stimulation were either left untreated or were incubated at 37°C with the combination of either PMA (20 ng/ml) (Sigma)/Ionomycin (Iono; 1 μM) (EMD Biosciences, San Diego CA) or anti-CD3 antibodies (clone OKT3 at 3 μg/ml) added to DT30 cells (10^5 DT30/10^5 target cells) in a final volume of 5 ml. Incubation with the various stimulating agents was terminated by the addition of ice-cold PBS, and nuclear extracts were prepared according to the described microscale preparation protocol [38]. Protein concentrations were determined by the bicinchoninic assay with a commercial protein reagent kit (Pierce, Rockford, IL). EMSA was performed with 10 μg of nuclear extracts incubated for 20 min at room temperature in 10 μl of 1X binding buffer (100 mM HEPES, pH 7.9, 40% glycerol, 10% Ficoll, 250 mM KCl, 10 mM dithiothreitol, 5 mM EDTA, 250 mM NaCl), 1 μg of poly(dI-dC), 5 μg of nuclease-free bovine serum albumin (fraction V) containing 0.8 ng of γ-^{32}P-labeled double-stranded DNA oligonucleotide. The tested probe was synthesized in-house and consisted of the distal NFAT-binding site from the murine IL-2 promoter (5'-TCGAGCCCAAAGAGGAAA ATTTGTTTCATG-3').

DNA-protein complexes were resolved from free labeled DNA by electrophoresis on a native 6% (w/v) polyacrylamide gel. Gels were subsequently dried and exposed on a Kodak™ Biomax MR film at -85°C.

Results

NL4-3 Viruses Preferentially Replicate in Memory CD4+ T Cells

Although preferential replication of HIV-1 in CD45RO CD4+ T lymphocyte has been previously demonstrated [7, 15-18, 20], the mechanism behind this observation needs further studies. To grasp a better understanding of the underlying mechanism behind this preferential replication, HIV-1 replication was directly tested in both naïve and memory CD4+ T cell populations. Using immobilized antibody-coated magnetic beads, the two CD4+ T cell populations were separated from purified PBMCs of healthy donors by negative selection, thus reducing the likelihood of biased measurements resulting from antibody-mediated cross-linking of molecules on the cell surface of the isolated population. Flow cytometric analyses demonstrated that sorted CD4+ T cell populations generally yielded a greater than 90% level of purity (data not shown). Both purified memory and naïve T cell subsets were used for subsequent experiments.

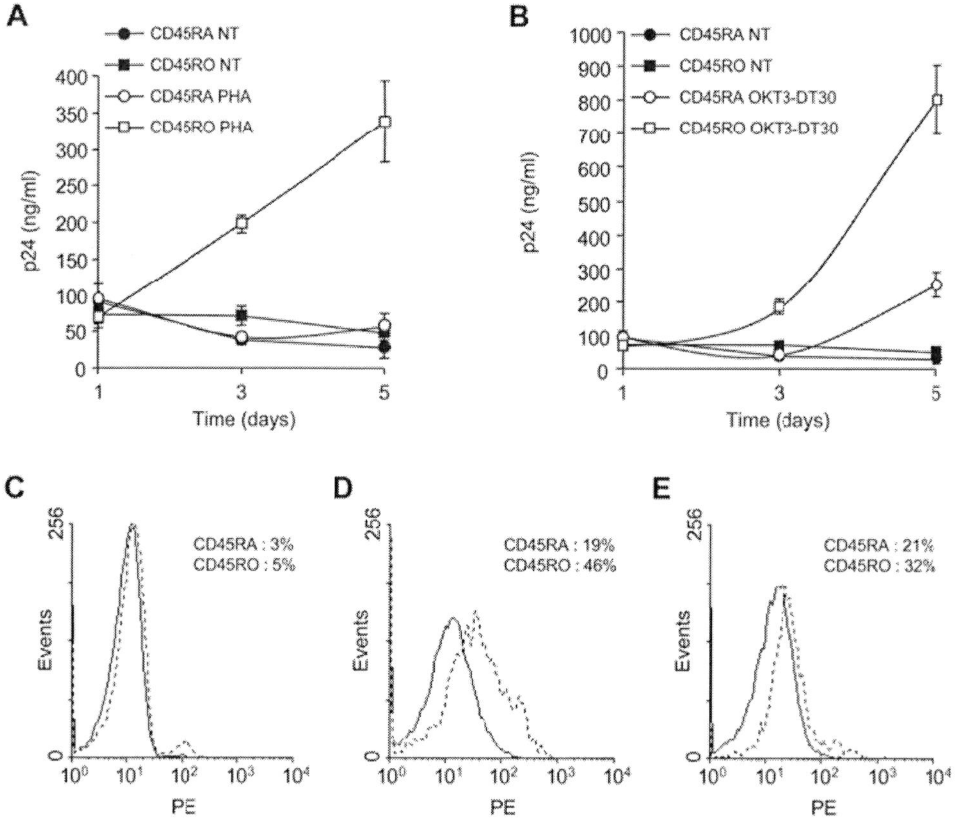

Figure_1. HIV-1 replicates more efficiently in memory CD4+ T cells. A, B. Purified CD45RA (circles) and CD45RO (squares) CD4+ T lymphocytes were infected with NL4-3 (0.5 pg of p24 per cell) for 16 hours and were either left untreated (filled symbols) or stimulated with PHA (3 μg/ml) (A) or OKT3-DT30 (3 μg/ml OKT3 with 10^5 DT30/10^5 target cells) (B) (empty symbols) for 5 days. Cell-free supernatants were collected after 1, 3, and 5 days. Diluted supernatants were spotted in 96-well plates, and p24 were determined by p24 ELISA. Each value represents the mean value ±S.D. calculated from three different cell samples in the same experimental setting. C to E. CD45RA (——) and CD45RO (------) CD4+ T lymphocytes were infected with NL4-3 (0.5 pg of p24 per cell) for 16 hours and were either left untreated (C) or incubated with PHA (3 μg/ml) (D) or OKT3-DT30 (3 μg/ml OKT3 with 10^5 DT30/10^5 target cells) (E) for 5 days. Intracellular p24 antigen was analyzed through flow cytometry using clone 31-90-25 (anti-p24) as the primary antibody and an (R)-PE-conjugated goat anti-IgG secondary antibody. The number of cells positive for the p24 antigen is determined by comparison to uninfected cells and is indicated in each graph as percentage of total number of cells for each cell population. These results are representative of two independent experiments.

Isolated CD4+CD45RA+ and CD4+CD45RO+ T lymphocytes were infected with equal amount of wild-type NL4-3 virions. Stimulation was then performed in the presence of PHA or DT30 cells (expressing the human B7.1 and the Fcγ receptor) incubated with OKT3 antibodies. This latter combination allowed us to mimic a typical antigen-dependent T-cell activation. Supernatant-associated p24 was quantified at days 1, 3 and 5 post-stimulation. As presented in Figure 1 (panels A and B), although limited amounts of viral p24 was detected in both unstimulated T cell subsets at day 1, no important differences were noted between the two cell subsets in these measured p24 levels suggesting equal susceptibility to the early steps of infection. However, at days 3 and 5, stimulation by either PHA or OKT3-DT30 induced significantly higher HIV-1 replication levels in memory CD4+ T cells when compared to similarly treated naïve CD4+ T cells.

In these infection experiments, the levels of intracellular p24 was also evaluated by intracellular FACS analysis at day 5 post-stimulation (Figure 1 panels C, D and E).

These results again indicated higher levels of intracellular p24 in memory CD4+ T cells upon activation with either OKT3-DT30 or PHA. These results hence confirmed previous studies showing higher HIV-1 replication level in the memory CD4+T cell subset upon stimulation.

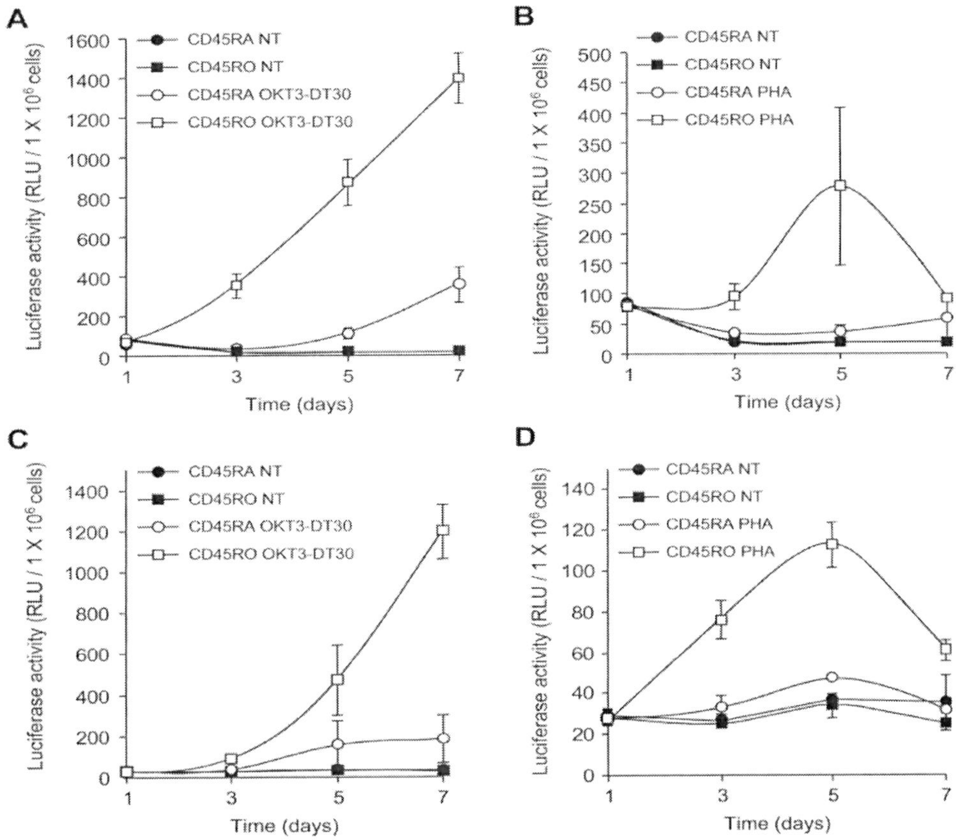

Figure_2. Greater levels of luciferase activity in memory CD4+ T cells infected with pseudotyped HIV-1 virions. Purified CD45RA (circle) and CD45RO (square) CD4+ T lymphocytes were infected with HXB2 envelope-pseudotyped NL4-3 Luc+Env- virions (A and B) or HXB-LUC (C and D) (0.5 pg of p24 per cell) for 16 hours and either left untreated (filled symbols) or stimulated with OKT3-DT30 (3 μg/ml OKT3 with 10^5 DT30/10^5 target cells) (A and C) or PHA (3 μg/ml) (B and D) (empty symbols). Cells were next harvested and lysed after 1, 3, 5 and 7 days of stimulation. Results are presented in terms of relative light units (RLU) per 1 x 10^6 cells from the calculated means ± S.D. of three different lysed cell samples in the same experimental setting. These results are representative of three different experiments.

Higher Replication of Luciferase-Expressing Pseudotyped HIV-1 Viruses in Memory CD4+ T Cells

To assess which step of the HIV-1 replication cycle was involved in this preferential replication of HIV-1 for memory CD4+ T cells, HXB-LUC virions, which express the luciferase reporter gene, were next used to infect purified CD45RA and CD45RO CD4+ T lymphocytes. Viral infection was also conducted with luciferase-expressing NL4-3 viruses deficient for the production of the envelope (NL4-3 Luc+Env-). These deficient viruses were thereby recomplemented with the HIV-1 envelope derived from the HXB-2 strain. Luciferase activity was measured at different time points following infection and is therefore representative of single round infection events.

Isolated memory and naïve CD4+ T cell populations were incubated with equal amounts of HXB-LUC and pseudotyped NL4-3 Luc+Env- virions and then stimulated with either OKT3-DT30 or PHA. As depicted in Figure 2, in unstimulated condition, the level of luciferase activity was similar for the two infected CD4+ T cell populations again suggesting equal susceptibility to infection toward these viruses. On the other hand, from day 3 to day 7, luciferase activity markedly increased in the memory T cell subset infected with both HIV-1 virions and stimulated with OKT3-DT30, while a limited increase was noted for naïve T cells (Figure 2, panels A and C). Similar results were obtained in PHA-stimulated T cell subsets upon infection by these virions, although, for memory T cells, HIV-1 replication was noted to rapidly decrease at day 7 (Figure 2, panels B and D). In addition, the induction of luciferase activity was generally low. These results nonetheless confirmed that HIV-1 replication is higher in memory CD4+ T cells and also indicated that this preferential replication occurred either at the level of gene expression or at earlier steps of the replicative cycle.

Preferential Replication of HIV-1 in CD4+CD45RO+ T Cells is Independent of Viral Entry

It had already been suggested that naïve and memory CD4+T cell subsets are equally susceptible to HIV-1 infection, although recent studies have suggested differences in viral entry between naïve and memory CD4+ T cells [21, 22]. The results presented in Figure 1 and 2 suggested that HIV-1 viral entry was equivalent in both T cell subsets, as equal supernatant-associated p24 and extract-associated luciferase activity were present in CD4+ T cells of both subsets in unstimulated conditions.

Nonetheless, to more appropriately determine their susceptibility toward HIV-1 infection at the level of viral entry, we measured intracellular p24 antigen levels in both T lymphocyte populations after a short time of exposure to the virus hence allowing detection of newly entered HIV-1 capsids.

The results obtained from this assay are presented in Figure 3 (panel A). After a brief incubation of both T cell subsets to pseudotyped NL4-3 Luc+Env- virions or with the wild type NL4-3 virus, results indicated that, regardless of the tested virus, naïve and memory CD4+ T cells contained comparable levels of intracellular p24 antigen strongly arguing for equal efficiency of viral entry.

We also tested infection susceptibility of both T cell populations toward NL4-3 Luc+Env- virions pseudotyped with a different envelope; i.e. VSV-G. Regardless of the change in tropism and mode of entry of this new pseudotyped HIV-1 virion, after stimulation with OKT3-DT30, memory CD4+ T cells again showed higher luciferase activity upon infection than in similarly treated naïve CD4+ T cells (Figure 3, panel B).

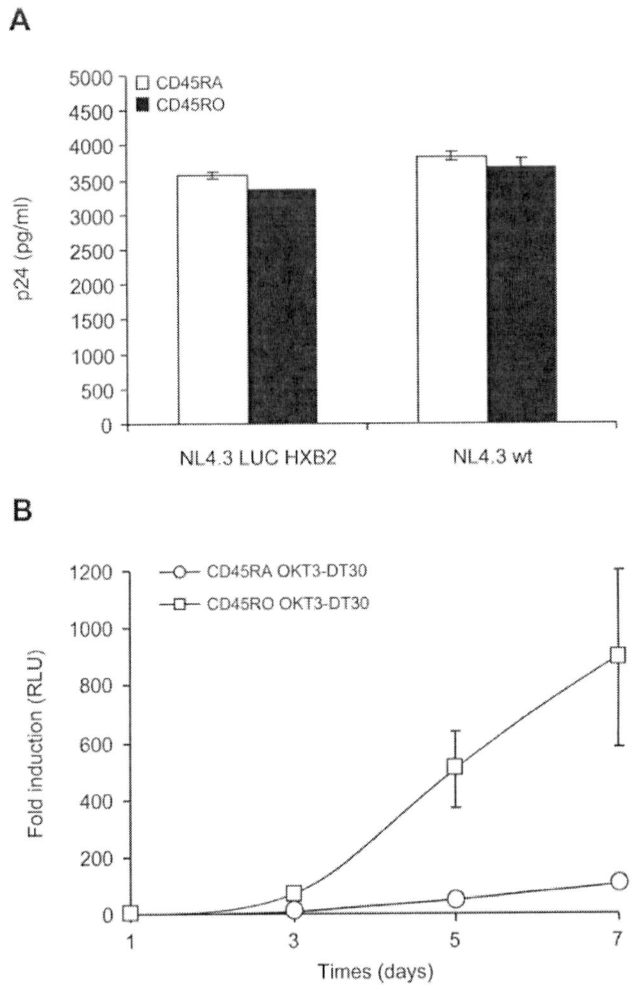

Figure 3. Preferential replication of HIV-1 in memory CD4+ T cells is viral entry-independent. (A) Purified CD45RA (empty columns) and CD45RO (black columns) CD4+ T cells were exposed to HXB2 envelope-pseudotyped NL4-3 Luc+Env- or NL4-3 viruses (0.5 pg of p24 per cell) for 1 h at 37°C. Trypsinized cells were lysed and evaluated for intracellular p24 content by the p24 ELISA assay. Each value represents the mean value ±S.D. calculated from three different cell samples in the same experimental setting. (B) Purified CD45RA (circle) and CD45RO (square) CD4+ T lymphocytes were infected with VSV-G envelope-pseudotyped NL4-3 Luc+Env- virions (0.5 pg of p24 per cell) for 16 hours and either left untreated or stimulated with OKT3-DT30 (3 μg/ml OKT3 with 10^5 DT30/10^5 target cells). Cells were collected and lysed after 1, 3, 5 and 7 days. Results are presented as fold inductions of luciferase activity from treated over untreated samples determined as the calculated means ± S.D. of three different lysed cell samples in the same experimental setting. These results are representative of two different experiments.

In addition, luciferase activity levels were equivalent throughout the experiments in unstimulated cells (data not shown). Furthermore, viral entry assay also indicated identical entry of the VSV-G-pseudotyped HIV-1 virions in both CD4+ T cell subsets (data not shown). The above results thus further indicated that preferential replication of HIV-1 in memory CD4+ T cells was viral entry-independent.

Higher Levels of Reverse Transcribed HIV-1 DNA in CD4+CD45RO+ T Cells Following Activation

After viral entry, HIV-1 RNA is subsequently reverse-transcribed into DNA, which then integrates into cellular DNA. These events of the HIV-1 replicative cycle are modulated by various cellular components [39] To confirm that differences in HIV-1 replication susceptibility between CD45RA vs. CD45RO CD4+ T cells were taken place after fusion, real-time PCR approach was optimized to evaluate changes in the content of viral DNA between the different cell subsets at 5 day postinfection.

Both LTR-associated M667 and AA55 primers were initially chosen as they allowed specific detection of HIV-1 DNA strand [40]. In parallel, the β-actin gene was amplified in order to normalize for variation in amplification efficiency in between samples. Total DNA was extracted from both T cell subsets infected by NL4-3 and induced by PHA or OKT3-DT30 and was thus analyzed by real-time PCR using both primers described above. As depicted in Figure 4 (panel A), activated CD45RA and CD45RO CD4+ T cells showed higher level of viral DNA than in the unstimulated cell DNA samples. However, these experiments clearly demonstrated that PHA-and OKT3-DT30-stimulated memory CD4+ T cells contained higher levels of viral DNA in comparison to similarly activated naïve CD4+ T cells (7 fold and 5 fold higher levels of viral DNA, respectively). These results hence suggested that, upon stimulation, more viral DNA could be detected in infected memory CD4+ T cells.

A more targeted approach was next undertaken to further confirm the observed differences in viral DNA content measured from the two CD4+ T cell populations. The previously described *Alu*-based real-time PCR method was used to quantify the levels of integrated proviral DNA [37]. This method specifically amplifies integrated DNA by an initial round of PCR amplification with the *Alu*-specific primer, which allows amplification of proviral DNA integrated at a nearby distance from this highly abundant human retroelement sequence. Real-time PCR was then performed with the PCR primers described in the above real-time PCR approach and the resulting signals were normalized against the amplified β-actin gene. Again, as demonstrated above (Figure 4, panel B), the viral copy number attributed to integrated viral DNA was higher in activated T cells (in comparison to their unstimulated counterpart).

However, noticeably, higher integrated viral DNA copies were noted in the HIV-1-infected activated CD45RO CD4+ T cell population in comparison to stimulated naïve T cells. These increases in the content of integrated HIV-1 DNA were noted for both PHA- and OKT3-DT30-activated T cells. In both sets of real-time PCR experiments, viral DNA was undetectable in uninfected cells. Gel analysis of the amplified signal at the end of real-time PCR experiments also indicated the presence of the amplified product at the expected size (data not shown). These results hence confirmed that differences in HIV-1 replication between memory and naïve CD4+ T cells are likely attributable to a step occurring at later time points in the replicative cycle of the virus.

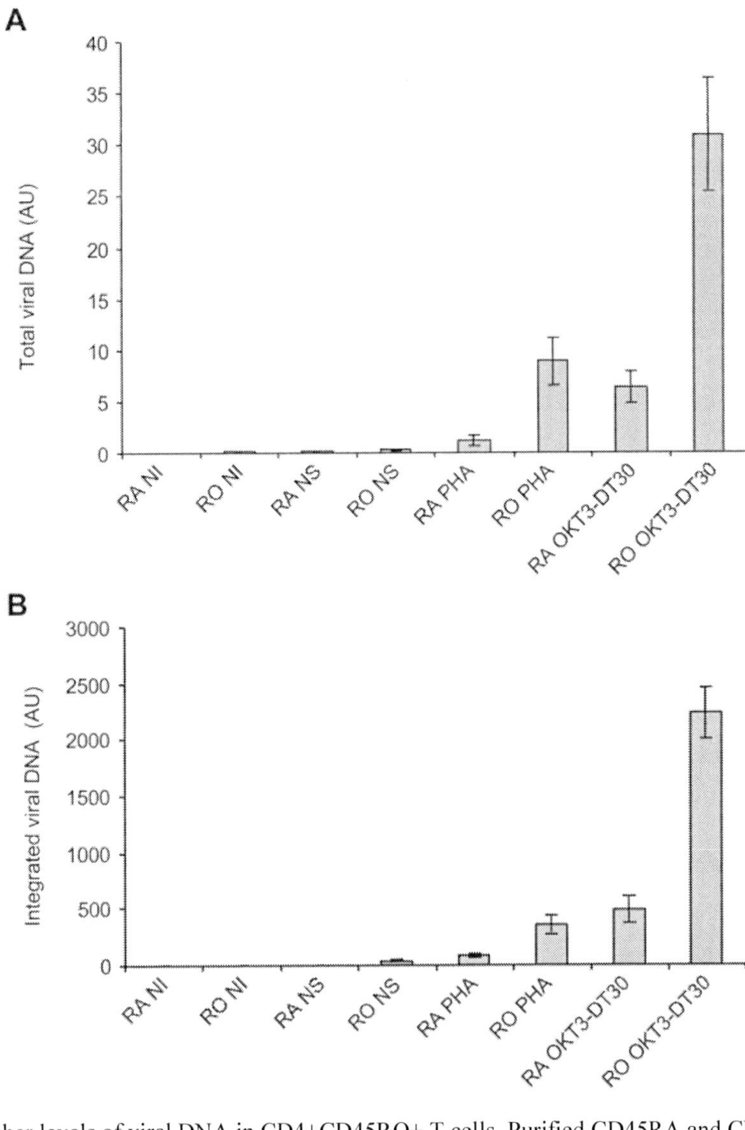

Figure 4. Higher levels of viral DNA in CD4+CD45RO+ T cells. Purified CD45RA and CD45RO CD4+ T cells were infected with NL4-3 (0.5 pg of p24 per cell) for 16 hours and left untreated (NS) or stimulated with PHA (3 µg/ml) or OKT3-DT30 (3 µg/ml OKT3 with 10^5 DT30/10^5 target cells). Total cellular DNA was extracted from uninfected cells (NI) and infected cells. (A) Amplification of the R/U5 segment from total HIV-1 viral DNA was conducted and quantified by real-time PCR using AA55 and M667 primers. (B) Integrated HIV-1 proviral DNA was selectively amplified with a first round of PCR with *Alu* and M661 primers. Integrated viral DNA was subsequently quantified from diluted samples of the first PCR reaction by real-time PCR using AA55 and M667 primers. The copy number of HIV-1 DNA was determined in reference to a standard curve generated by serial dilution of pNL4-3 DNA. Amplification of a segment of the human β-actin gene served as a control for normalization. These results are representative of three different experiments.

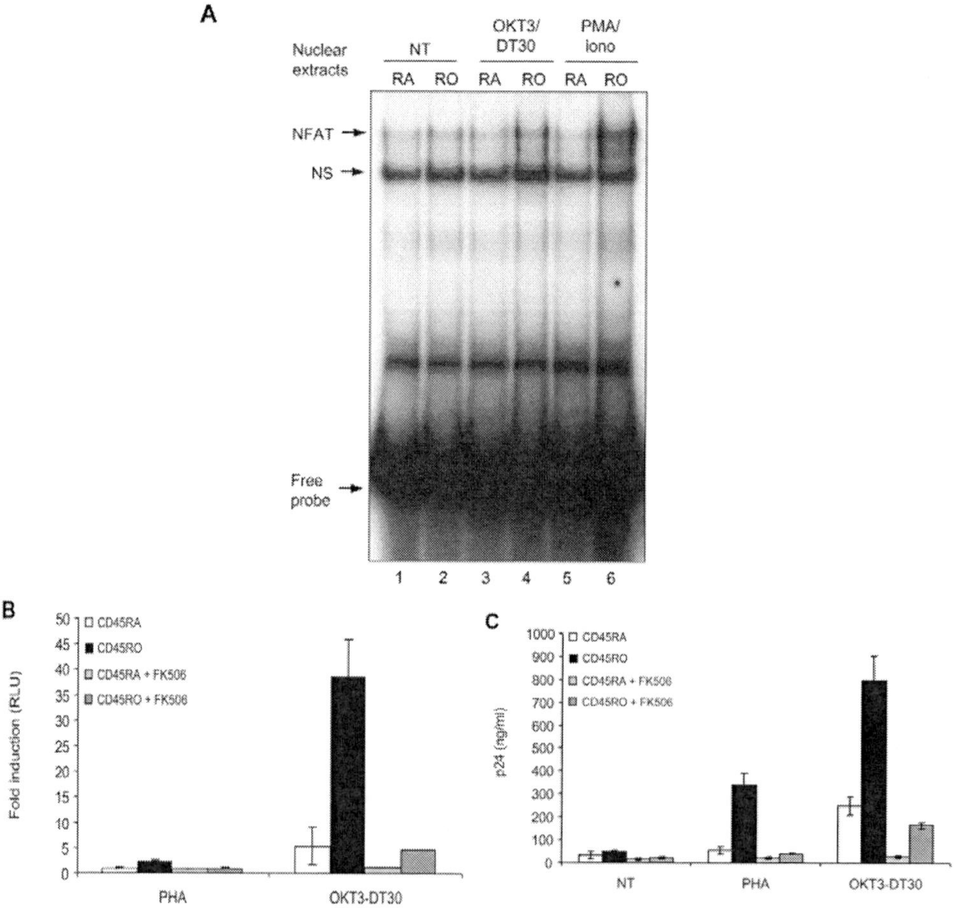

Figure 5. NFAT is more importantly activated in memory CD4+ T cells and preferential HIV-1 replication is sensitive to FK506. (A) Nuclear extracts from purified CD45RA and CD45RO CD4+ T cells either left untreated (lanes 1 and 2) or stimulated with OKT3-DT30 (3 μg/ml OKT3 with 10^5 DT30/10^5 target cells) (lanes 3 and 4) or with PMA (20 ng/ml)/Iono (1 μM) (lanes 5 and 6) were incubated with a NFAT probe. Protein/DNA complexes were analyzed on a 6% native polyacrylamide gel, which was dried and exposed on Kodak X OMAT film. The arrows indicate NFAT-specific complexes as well as non-specific complexes (NS) and free probes. (B) Purified CD45RA and CD45RO CD4+ T cells were infected with HXB-LUC (0.5 pg of p24 per cell) for 16 hours and subsequently treated or not with FK506 (10 ng/ml) for 1 hour. Infected cells were then left unstimulated or stimulated with PHA (3 μg/ml) or OKT3-DT30 (3 μg/ml OKT3 with 10^5 DT30/10^5 target cells). Cells were collected and lysed after 7 days. Results are presented as fold inductions of luciferase activity from treated over untreated sample determined as the calculated means ± S.D. of three different lysed cell samples in the same experimental setting. (C) Purified CD45RA and CD45RO CD4+ T cells were infected with NL4-3 (0.5 pg of p24 per cell) for 16 hours and subsequently treated or not with FK506 (10 ng/ml) for 1 hour. Infected cells were then left unstimulated or stimulated with PHA (3 μg/ml) or OKT3-DT30 (3 μg/ml OKT3 with 10^5 DT30/10^5 target cells). Cell-free supernatants were collected after 5 days. Diluted supernatants were spotted in 96-well plates, and p24 were determined by p24 ELISA. Each value represents the mean value ±S.D. calculated from three different cell samples in the same experimental setting. These results are representative of two different experiments.

Role of NFAT in Preferential HIV-1 Replication Observed in Memory CD4+ T Cells

The activation of T lymphocytes induces a series of signaling transduction events leading to activation of multiple transcription factors including NFAT. This transcription factor has been previously demonstrated to play an active role in both HIV-1 gene expression and reverse transcription, making it a possible candidate contributing to preferential HIV-1 replication in memory CD4+ T cells. To test this possibility, we first conducted EMSA experiments on nuclear extracts isolated from memory and naïve CD4+ T cells, which had been stimulated with two combinations of activation agents known to potently activate NFAT. These extracts were then incubated with a probe consisting of a consensus NFAT-binding site. As depicted in Figure 5 (panel A), extracts from OKT3-DT30- or PMA/Ionomycin-activated CD45RO CD4+ T cells indeed presented a stronger NFAT-specific signal than in the stimulated naïve T cell population (lanes 3 versus 4 and 5 versus 6). The results presented in these analyses hence concurred toward the possible implication of NFAT in the difference in HIV-1 replication efficacy between the two CD4+ T cell subsets.

To further underscore the importance of NFAT in the preferential replication of HIV-1 in memory CD4+ T cells, FK506, an NFAT inhibitor, was added at subcytotoxic concentrations to both isolated memory and naïve CD4+ T lymphocytes after infection with either HXB-LUC or NL4-3 viruses (Figure 5, panels B and C respectively). Following activation, viral replication was either assessed for luciferase activity (Panel B) or for levels in supernatant-associated p24 antigen (panel C).

As demonstrated in these panels, both viruses again replicated at higher levels in memory CD4+ T cells in PHA- as well as OKT3-DT30-stimulating conditions. However, the addition of FK506 to both T cell populations had an important effect on the difference in HIV-1 replication susceptibility. Indeed, measurement of viral replication for HXB-LUC and NL4-3 demonstrated that FK506 greatly affected luciferase activity and p24 antigen levels at a more important level for memory CD4+ T cells rendering both cell types poorly suitable cell environment for HIV-1 replication. Similar experiments were conducted on infected PMA/Ionomycin-stimulated CD4+ T cells and also led to similar conclusion in that FK506 strongly diminished the difference in HIV-1 replication observed between both CD4+ T cell populations (data not shown).

These results strongly argue that the NFAT transcription factor is a key element for the high level of HIV-1 replication occurring in memory CD4+ T lymphocytes and could alter the efficiency of reverse transcription and the extent of HIV-1 gene expression.

CONCLUSION

Naïve and memory T cells differ in the expression of the various CD45 isoforms and the levels of cell surface adhesion molecules. Naïve T cells express the CD45RA isoform and very low level of activation and adhesion molecules on their surface, whereas memory T cells are CD45RO-positive and express high levels of cell surface adhesion molecules, possibly reflecting their lower requirement for costimuli during TCR activation [41, 42]. Former studies had established that HIV-1 preferentially replicates in memory T cells, although the exact mechanism leading to this difference in replication efficacy in the two cell types was not clearly outlined [7, 15-20]. Using a Jurkat cell line model, we had earlier designated NFAT as a key element for this preferential HIV-1 replication observed in CD45RO-expressing T cells (16). In the present study, using isolated human memory and naïve CD4+ T cells, we demonstrate that a post-fusion step in the HIV-1 life cycle is more efficient in memory CD4+ T cells and that NFAT is likely at the basis for this difference in efficiency.

Memory and naïve CD4+ T cell subsets were isolated by negative selection with immunomagnetic beads and were first infected with NL4-3 viruses. By measuring viral p24 antigen in the supernatant and in infected cells, greater levels of viral production were detected in memory CD4+ T cells following T cell activation. This difference was noted using the antigenic-like OKT3-DT30-stimulating condition thereby corroborating previous studies [7, 15-20]. However, previous results had inferred that PHA stimulation did not lead to preferential replication in memory CD4+ T cells [15, 43, 44], which is in contrast with our results. It is interesting to note that, similar to the study of Medina *et al.* [45], we have also observed a faster and more transient peak in HIV-1 replication for PHA-stimulated cells in comparison to OKT3-DT30-stimulated cells when luciferase-expressing HIV-1 virions were tested. Variation in the time kinetic of induced HIV-1 replication by these two different activating agents could explain why previous studies have not seen differences in levels of HIV-1 replication between both PHA-activated CD4+ T cell subsets.

We have also observed a difference in luciferase reporter gene expression when memory and naïve CD4+ T lymphocytes were infected with luciferase-expressing HIV-1 virions. In accordance to our results, viral replication was more intense in memory CD4+ T cells following PHA or OKT3-DT30 treatment. These results thus argued that the difference in HIV-1 replication was a consequence of a difference of efficiency in replication steps occurring at or before activation of HIV-1 gene expression. Previous studies had suggested that memory and naïve CD4+ T cells were equally susceptible to infection of HIV-1 [18]. Through a viral entry assay, we have confirmed these data. In addition, VSV-G-pseudotyped HIV-1 virions also demonstrated higher HIV-1 replication in memory CD4+ T cells after cell activation, again indicating that viral entry was not a determining factor in the preferential HIV-1 replication occurring in memory CD4+ T cells.

In order to determine the steps involved in the high level of viral replication in memory CD4+ T cells, we then conducted different real-time PCR experiments. Analysis of these results allowed us to conclude that, in PHA- and-OKT3-DT30-activated T cells, a higher viral DNA copy number was present in infected memory CD4+ T cells in both total and integrated viral DNA again arguing for a more efficient infection of memory CD4+ T cells. These results differ from the data of Spina *et al.* [15], which demonstrated equivalent levels of viral DNA in both phenotypic CD4+ T subsets. As standard PCR was used for their analyses, these data might not have properly assessed potential changes in viral DNA and we are hence more confident in the data generated from real-time PCR, which is a more quantitative measurement of viral DNA levels. In fact, our results agree with the study of Ostrowski *et al,* which had previously indicated more abundant HIV-1 proviral integrated in memory CD4+ T cells of infected patients [46].

As NFAT had been demonstrated to regulate the retrotranscription process of HIV-1 [47], this transcription factor was selected as a potential candidate for the observed difference in HIV-1 replication between both studied CD4+ T cell populations. Through EMSA analysis, we have indeed demonstrated that NFAT was more abundant in nuclear extracts from memory CD4+ T cells than in naïve CD4+ T cells activated by either OKT3-DT30 or PMA/Ionomycin. Upon FK506 treatment, our results further argued for an involvement of NFAT in the high levels of HIV-1 replication observed in memory CD4+T cells as this NFAT inhibitor importantly reduced differences in luciferase activity and p24 antigen levels between the two CD4+ T cell subsets. Our results thus further underscore the important role played by NFAT in HIV-1 replication. More specifically, we would argue that NFAT, through its previously reported role on retrotranscription, permits higher levels of HIV-1 replication in memory CD4+ T cells. In fact, our results parallel those presented by Ofori *et al.* [48], who provided a similar explanation for higher HIV-1 replication observed in Th2 cells in comparison to Th1 cells. It should be stressed however that we cannot exclude a possible increase in both the integration process and activation of gene expression in the observed higher level of replication of HIV-1 in memory CD4+ T cells. In fact, through the Jurkat cell line model, we had previously demonstrated that higher HIV-1 replication in CD45RO-expressing Jurkat cell lines was NFAT-dependent and paralleled higher levels of LTR activation in these cells [19]. We have conducted real-time RT-PCR experiments and have indeed demonstrated higher levels of viral gene expression in activated memory CD4+ T cells (B.A. and B.B., unpublished results), although these increases in gene expression are likely a result of higher levels of viral DNA in infected memory CD4+ T cells. In addition, previous reports using a *SCID*-hu (Thy/Liv) mouse model or a novel latency model based on isolated primary CD4+ T cells have indicated that the activation of NFAT was important for the activation of latent HIV-1[49, 50]. More experiments are required to determine whether these other steps in HIV-1 replication could indeed impart a more active HIV-1 replication in memory CD4+ T cells.

Although our data suggest an important involvement of NFAT on HIV-1 replication in memory CD4+ T cells, EMSA analysis indicate that NF-κB is more importantly activated in memory CD4+ T cells and thus might also have a potential influence (B.A. and B.B., unpublished results). In fact, the involvement of NF-κB has also been indirectly suggested in experiments conducted on naïve and memory CD4+ T cells treated with PMA alone, again demonstrating higher replication levels in stimulated memory CD4+ T cells infected with luciferase-expressing HIV-1 virions (data not shown).

Our results thus strongly suggest that preferential replication of HIV-1 in memory CD4+ T lymphocytes is mediated by a more active process of reverse transcription. Our data also provide a rationale to include NFAT as potential targets for antiretroviral therapies. However, many steps of the HIV-1 life cycle beside reverse transcription might be involved in this high level of permissiveness of memory CD4+ T cells for HIV-1 replication and experiments are presently underway to further address these issues.

ACKNOWLEDGEMENTS

We thank Dr. M. Dufour for excellent technical assistance in flow cytometry studies. B.B. was supported by a scholarship (Junior 2 level) from the Fonds de la Recherche en Santé du Québec (FRSQ). M.J.T. is the recipient of a Senior Canada Research Chair in Human Immuno-Retrovirology. This work was supported by a grant to B.B. from the Canadian Foundation for AIDS Research (CANFAR) (grant #014 032).

References

[1] Klatzmann D, Champagne E, Chamaret S, Gruest J, Guetard D, Hercend T, Gluckman JC and Montagnier L. T-lymphocyte T4 molecule behaves as the receptor for human retrovirus LAV. *Nature*, 1984;312(5996):767-8.

[2] Lasky LA, Nakamura G, Smith DH, Fennie C, Shimasaki C, Patzer E, Berman P, Gregory T and Capon DJ. Delineation of a region of the human immunodeficiency virus type 1 gp120 glycoprotein critical for interaction with the CD4 receptor. *Cell.*, 1987;50(6):975-85.

[3] Moore JP, McKeating JA, Norton WA and Sattentau QJ. Direct measurement of soluble CD4 binding to human immunodeficiency virus type 1 virions: gp120 dissociation and its implications for virus-cell binding and fusion reactions and their neutralization by soluble CD4. *J. Virol.*, 1991;65(3):1133-40.

[4] Fortin JF, Barbeau, B., Robichaud, G.A., and Tremblay, M.J. T cell activation and regulation of HIV-1 : same effectors with distinct outcomes. *Canadian Journal of Infectious Diseases,* 1999;10(Suppl. C):25C-32C.

[5] Trowbridge IS and Thomas ML. CD45: an emerging role as a protein tyrosine phosphatase required for lymphocyte activation and development. *Annu. Rev. Immunol.,* 1994;12:85-116.

[6] Thomas ML. The leukocyte common antigen family. *Annu. Rev. Immunol.,* 1989;7:339-69.

[7] Justement LB. The role of CD45 in signal transduction. *Adv. Immunol.,* 1997;66:1-65.

[8] Berard M and Tough DF. Qualitative differences between naive and memory T cells. *Immunology*, 2002;106(2):127-38.

[9] Merkenschlager M, Terry L, Edwards R and Beverley PC. Limiting dilution analysis of proliferative responses in human lymphocyte populations defined by the monoclonal antibody UCHL1: implications for differential CD45 expression in T cell memory formation. *Eur. J. Immunol.,* 1988;18(11):1653-61.

[10] Leitenberg D, Novak TJ, Farber D, Smith BR and Bottomly K. The extracellular domain of CD45 controls association with the CD4-T cell receptor complex and the response to antigen-specific stimulation. *J. Exp. Med.,* 1996;183(1):249-59.

[11] Onodera H, Motto DG, Koretzky GA and Rothstein DM. Differential regulation of activation-induced tyrosine phosphorylation and recruitment of SLP-76 to Vav by distinct isoforms of the CD45 protein-tyrosine phosphatase. *J. Biol. Chem.,* 1996;271(36):22225-30.

[12] Hall SR, Heffernan BM, Thompson NT and Rowan WC. CD4+ CD45RA+ and CD4+ CD45RO+ T cells differ in their TCR-associated signaling responses. *Eur. J. Immunol.,* 1999;29(7):2098-106.

[13] Lassen K, Han Y, Zhou Y, Siliciano J and Siliciano RF. The multifactorial nature of HIV-1 latency. *Trends. Mol. Med.,* 2004;10(11):525-31.

[14] Han Y, Wind-Rotolo M, Yang HC, Siliciano JD and Siliciano RF. Experimental approaches to the study of HIV-1 latency. *Nat. Rev. Microbiol.,* 2007;5(2):95-106.

[15] Spina CA, Prince HE and Richman DD. Preferential replication of HIV-1 in the CD45RO memory cell subset of primary CD4 lymphocytes in vitro. *J. Clin. Invest.,* 1997;99(7):1774-85.

[16] Woods TC, Roberts BD, Butera ST and Folks TM. Loss of inducible virus in CD45RA naive cells after human immunodeficiency virus-1 entry accounts for preferential viral replication in CD45RO memory cells. *Blood,* 1997;89(5):1635-41.

[17] Tsunetsugu-Yokota Y, Kato T, Yasuda S, Matsuda Z, Suzuki Y, Koyanagi Y, Yamamoto N, Akagawa K, Cho MW and Takemori T. Transcriptional regulation of HIV-1 LTR during antigen-dependent activation of primary T cells by dendritic cells. *J. Leukoc. Biol.,* 2000;67(3):432-40.

[18] Roederer M, Raju PA, Mitra DK and Herzenberg LA. HIV does not replicate in naive CD4 T cells stimulated with CD3/CD28. *J. Clin. Invest.,* 1997;99(7):1555-64.

[19] Robichaud GA, Barbeau B, Fortin JF, Rothstein DM and Tremblay MJ. Nuclear factor of activated T cells is a driving force for preferential productive HIV-1 infection of CD45RO-expressing CD4+ T cells. *J. Biol. Chem.,* 2002;277(26):23733-41.

[20] Cayota A, Vuillier F, Scott-Algara D and Dighiero G. Preferential replication of HIV-1 in memory CD4+ subpopulation. *Lancet,* 1990;336(8720):941.

[21] Dai J, Agosto LM, Baytop C, Yu JJ, Pace MJ, Liszewski MK and O'Doherty U. Human immunodeficiency virus integrates directly into naive resting CD4+ T cells but enters naive cells less efficiently than memory cells. *J. Virol.,* 2009;83(9):4528-37.

[22] Tardif MR and Tremblay MJ. LFA-1 is a key determinant for preferential infection of memory CD4+ T cells by human immunodeficiency virus type 1. *J. Virol.,* 2005;79(21):13714-24.

[23] Kiani A, Rao A and Aramburu J. Manipulating immune responses with immunosuppressive agents that target NFAT. *Immunity,* 2000;12(4):359-72.

[24] Friedman J and Weissman I. Two cytoplasmic candidates for immunophilin action are revealed by affinity for a new cyclophilin: one in the presence and one in the absence of CsA. *Cell,* 1991;66(4):799-806.

[25] Schreiber SL and Crabtree GR. The mechanism of action of cyclosporin A and FK506. *Immunol Today,* 1992;13(4):136-42.

[26] Kinoshita S, Su L, Amano M, Timmerman LA, Kaneshima H and Nolan GP. The T cell activation factor NF-ATc positively regulates HIV-1 replication and gene expression in T cells. *Immunity,* 1997;6(3):235-44.

[27] Fortin JF, Barbeau B, Robichaud GA, Pare ME, Lemieux AM and Tremblay MJ. Regulation of nuclear factor of activated T cells by phosphotyrosyl-specific phosphatase activity: a positive effect on HIV-1 long terminal repeat-driven transcription and a possible implication of SHP-1. *Blood,* 2001;97(8):2390-400.

[28] Cron RQ, Bartz SR, Clausell A, Bort SJ, Klebanoff SJ and Lewis DB. NFAT1 enhances HIV-1 gene expression in primary human CD4 T cells. *Clin. Immunol,* 2000;94(3):179-91.

[29] Azuma M, Cayabyab M, Buck D, Phillips JH and Lanier LL. CD28 interaction with B7 costimulates primary allogeneic proliferative responses and cytotoxicity mediated by small, resting T lymphocytes. *J. Exp. Med.,* 1992;175(2):353-60.

[30] Deng H, Liu R, Ellmeier W, Choe S, Unutmaz D, Burkhart M, Di Marzio P, Marmon S, Sutton RE, Hill CM, Davis CB, Peiper SC, Schall TJ, Littman DR and Landau NR. Identification of a major co-receptor for primary isolates of HIV-1. *Nature,* 1996;381(6584):661-6.

[31] Yee JK, Friedmann T and Burns JC. Generation of high-titer pseudotyped retroviral vectors with very broad host range. *Methods Cell Biol.,* 1994;43 Pt A:99-112.

[32] Chen BK, Saksela K, Andino R and Baltimore D. Distinct modes of human immunodeficiency virus type 1 proviral latency revealed by superinfection of nonproductively infected cell lines with recombinant luciferase-encoding viruses. *J. Virol.,* 1994;68(2):654-60.

[33] Cantin R, Fortin JF and Tremblay M. The amount of host HLA-DR proteins acquired by HIV-1 is virus strain- and cell type-specific. *Virology,* 1996;218(2):372-81.

[34] Fortin JF, Cantin R, Lamontagne G and Tremblay M. Host-derived ICAM-1 glycoproteins incorporated on human immunodeficiency virus type 1 are biologically active and enhance viral infectivity. *J. Virol.,* 1997;71(5):3588-96.

[35] Cantin R, Martin G and Tremblay MJ. A novel virus capture assay reveals a differential acquisition of host HLA-DR by clinical isolates of human immunodeficiency virus type 1 expanded in primary human cells depending on the nature of producing cells and the donor source. *J. Gen. Virol.,* 2001;82(Pt 12):2979-87.

[36] Barbeau B, Bernier R, Dumais N, Briand G, Olivier M, Faure R, Posner BI and Tremblay M. Activation of HIV-1 long terminal repeat transcription and virus replication via NF-kappaB-dependent and -independent pathways by potent phosphotyrosine phosphatase inhibitors, the peroxovanadium compounds. *J. Biol. Chem.,* 1997;272(20):12968-77.

[37] Suzuki Y, Misawa N, Sato C, Ebina H, Masuda T, Yamamoto N and Koyanagi Y. Quantitative analysis of human immunodeficiency virus type 1 DNA dynamics by real-time PCR: integration efficiency in stimulated and unstimulated peripheral blood mononuclear cells. *Virus. Genes.,* 2003;27(2):177-88.

[38] Schreiber E, Matthias P, Muller MM and Schaffner W. Rapid detection of octamer binding proteins with 'mini-extracts', prepared from a small number of cells. *Nucleic Acids Res.,* 1989;17(15):6419.

[39] Frankel AD and Young JA. HIV-1: fifteen proteins and an RNA. *Annu. Rev. Biochem.,* 1998;67:1-25.

[40] Zack JA, Arrigo SJ, Weitsman SR, Go AS, Haislip A and Chen IS. HIV-1 entry into quiescent primary lymphocytes: molecular analysis reveals a labile, latent viral structure. *Cell,* 1990;61(2):213-22.

[41] Beverley P. Immunological memory in T cells. *Curr. Opin. Immunol.,* 1991;3(3):355-60.

[42] [42] Beverley PC. CD45 isoform expression: implications for recirculation of naive and memory cells. *Immunol. Res.,* 1991;10(3-4):196-8.

[43] Schnittman SM, Lane HC, Greenhouse J, Justement JS, Baseler M and Fauci AS. Preferential infection of CD4+ memory T cells by human immunodeficiency virus type 1: evidence for a role in the selective T-cell functional defects observed in infected individuals. *Proc. Natl. Acad. Sci. USA.,* 1990;87(16):6058-62.

[44] Janossy G, Borthwick N, Lomnitzer R, Medina E, Squire SB, Phillips AN, Lipman M, Johnson MA, Lee C and Bofill M. Lymphocyte activation in HIV-1 infection. I. Predominant proliferative defects among CD45R0+ cells of the CD4 and CD8 lineages. *AIDS.* 1993;7(5):613-24.

[45] Medina E, Borthwick N, Johnson MA, Miller S and Bofill M. Flow cytometric analysis of the stimulatory response of T cell subsets from normal and HIV-1+ individuals to various mitogenic stimuli in vitro. *Clin. Exp. Immunol.,* 1994;97(2):266-72.

[46] Ostrowski MA, Chun TW, Justement SJ, Motola I, Spinelli MA, Adelsberger J, Ehler LA, Mizell SB, Hallahan CW and Fauci AS. Both memory and CD45RA+/CD62L+ naive CD4(+) T cells are infected in human immunodeficiency virus type 1-infected individuals. *J. Virol.*, 1999;73(8):6430-5.

[47] Kinoshita S, Chen BK, Kaneshima H and Nolan GP. Host control of HIV-1 parasitism in T cells by the nuclear factor of activated T cells. *Cell*, 1998;95(5):595-604.

[48] Ofori H, Prokop J and Jagodzinski PP. Replication at the level of reverse transcription of HIV-1SF2 strain in Th1 and Th2 cells. *Biomed. Pharmacother.*, 2003;57(1):15-9.

[49] Brooks DG, Arlen PA, Gao L, Kitchen CM and Zack JA. Identification of T cell-signaling pathways that stimulate latent HIV in primary cells. *Proc. Natl. Acad. Sci. USA.*, 2003;100(22):12955-60.

[50] Bosque A and Planelles V. Induction of HIV-1 latency and reactivation in primary memory CD4+ T cells. *Blood*, 2009;113(1):58-65.

In: Tyrosine and Aspartic Acid
Editors: J. E. Jones and D. M. Morano
ISBN: 978-1-62100-752-4

Chapter 4

ASPARTIC ACID RACEMIZATION IN THE STUDY OF ANCIENT DNA IN BONES

Piera Iuliani and Lucia Di Federico
Dipartimento di Scienze del Farmaco,
Facoltà di Farmacia, Università degli Studi,
"G. D'Annunzio" di Chieti-Pescara, Via dei Vestini, Chieti, Italy

ABSTRACT

Only a small proportion of old specimens contain DNA that can be amplified by PCR: in the study of ancient DNA, reliable methods for the identification of the large fraction of samples that are so badly preserved that there is no need to attempt DNA extraction, especially in bones and teeth, are very useful. In particular, the extent of racemization of some amino acids has proved to be a very useful proxy for DNA preservation.

Almost all amino acids can occur in two optical isomers, termed L and D. Most naturally occurring proteins contain only L-amino acids, although there are some exceptional peptides which are biologically synthesized using D-form amino acids. Metabolically active tissues contain specialized enzymes known as racemases which maintain a disequilibrium in our cells of only the L isomers. Once an organism dies and its enzymes cease to function, amino acids undergo racemization, that is, the interconversion of L amino acids to D amino acids.

Racemization is affected by some of the factors that affect degradation of DNA. In particular, the rate of aspartic acid racemization is nearly identical to the rate of DNA depurination: in general, no DNA sequences of sufficient length can be extracted from samples in which the D/L Asp ratio is higher than 0.1.

In this chapter we review the major published methods in this field, especially referring to sample preparation, protein hydrolysis and D/L Asp ratio determination in fossil bones.

1. Introduction

Approximately the 99,9% of the over 50 billions of animal species which have populated the Earth throughout the geological time are extinct [1]. The rests of some of them, hominids and animals, are kept in museums. The study of DNA extracted from these fossils can give essential information on the genetic past of extinct animals, providing a useful tool to answer evolutionary and ecological questions. The development of Polymerase Chain Reaction (PCR) made it possible to study the ancient DNA in a scientifically rigorous way, allowing to obtain an almost unlimited number of copies of a single original DNA sequence [2-4]. The limitations inherent in this technique are substantially two: the molecular damage and the contamination.

After death, cellular DNA undergoes rapid degradation by lysosomial nucleases and at the same time suffers the aggression of microorganisms and insects [5]. In favorable conditions (such as the rapid drying of a tissue, the very low temperatures, the high saline concentration, the DNA absorption on a mineral matrix) the activity of nucleases and microorganisms is stopped before all nucleic acids are completely degradated. In these cases, slow but inexorable chemical processes start to affect DNA (oxidation, depurination and other hydrolytic processes), with consequent unescapable loss of information in the nucleotidic sequence [6]. After a sufficient period of time (which varies in function of environmental conditions), the hydrolytic damage is so wide that no amplifiable nucleic acids are still in the fossil sample. Moreover, depurination of some bases, even allowing amplification, causes the incorrect bases incorporation during PCR [2]. These miscoding lesions prevent the determination of the right sequence of the ancient DNA [7].

Another problem in the ancient DNA amplification is the high probability of contamination, because contemporary DNA is pervasive in the environment, both inside and outside the laboratory. This problem is obviously more serious in the study of human remains.

Therefore, rapid screening methods are crucial to identify the large fraction of samples which are unsuitable for DNA extraction. Amino acids analysis, through the determination of the total amount of amino acids preserved in a specimen, the composition in amino acids and the extent of their racemization, has proved to be a useful tool in the determination of DNA preservation and in supporting the authenticity of DNA sequences extracted from a specimen [6, 8-10]. Amino acids analysis is also widely applicable because the bulk of most fossilized tissues such as bone, skin and teeth is formed by proteins, that are usually in high concentration per gram of tissue and can be easily analysed by many current techniques [11]. Moreover proteins are good indicators of DNA preservation because proteic fraction of bone undergoes the same environmental aggression (temperature, humidity, etc.) of DNA.

2. Amino Acids Racemization

Proteins in living organisms are formed almost exclusively by L amino acids and their racemization is prevented by specialized enzymes which maintain a disequilibrium of the only L isomer. After death these enzymes cease to work and the two forms, D and L, start to interconvert. After a period of time adequately long, the D/L ratio approaches to one [8,11].

Racemization is a chemical reaction and several are the factors that influence its rate [12] such as amino acids structure, amino acidic sequence, pH, buffering effects, temperature, presence of water and/or of metallic cations. Racemization mechanism and kinetics are different between free and bound amino acids [13-17].

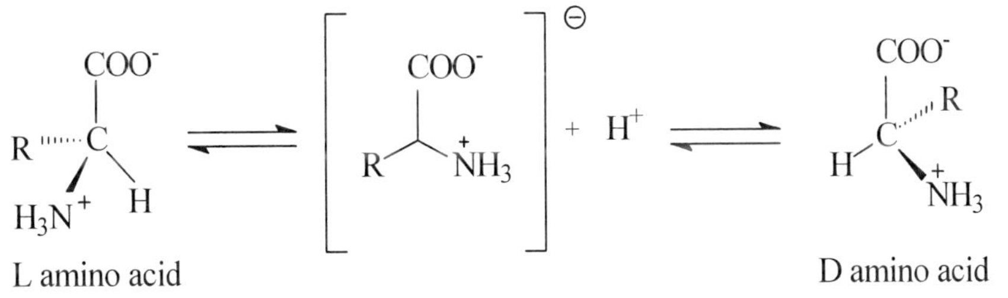

Figure 1. The racemization of a general free amino acid.

2.1. Free Amino Acids Racemization

In aqueous solution, racemization occurs via carbanion intermediate [18]: the removal of the hydrogen bonded to the amino acid α-carbon leads to the formation of a negatively charged planar carbanion. The readdition of a hydrogen ion to this intermediate has the same probability to generate the D or the L isomer (Figure 1).

Racemization reaction of amino acids in aqueous solution follows a first order kinetics, whose rate is influenced by:

1) Molecular structure: the presence on the α-carbon of groups able to stabilize the carbanion charge accelerate the racemization, contrarily groups which destabilize this charge retard it [17]. The overall effect of an R group depends on the combination of its electron withdrawing or donating properties (electronic effects) and its bulk and shape (steric factors), because steric hindrance interfere with proton removal, slowing the reaction. The rate racemization order for free amino acids in aqueous solution is: Ser > Thr > Asp > Phe > Ala > Glu > Leu > Ile > Val [17].
2) pH and buffering effects: the influence of pH on racemization rate strongly depends on temperature [16,19-20]. At 25°C racemization rate is independent from pH in the range from pH 3 to pH 9. Since most natural environments have a pH between 5 and 8, at the temperatures commonly found in nature the racemization rate is effectively independent from pH. Additionally, in biominerals such as bone, pH effects are minimized due to the buffering effect of hydroxylapatite matrix.
3) Temperature: in general, reaction rates increase with the increasing of temperature [21]. Free amino acids racemization follows a reversible first order kinetics [19, 22-23], so for a given temperature, reaction will proceed at a constant logarithmic rate until the reactant is consumed. Experiments have been carried out [22,24] to predict low temperature rates of racemization for some free amino acids (including aspartic acid) by extrapolation from high temperature data and to determine Arrhenius parameters for this reaction.

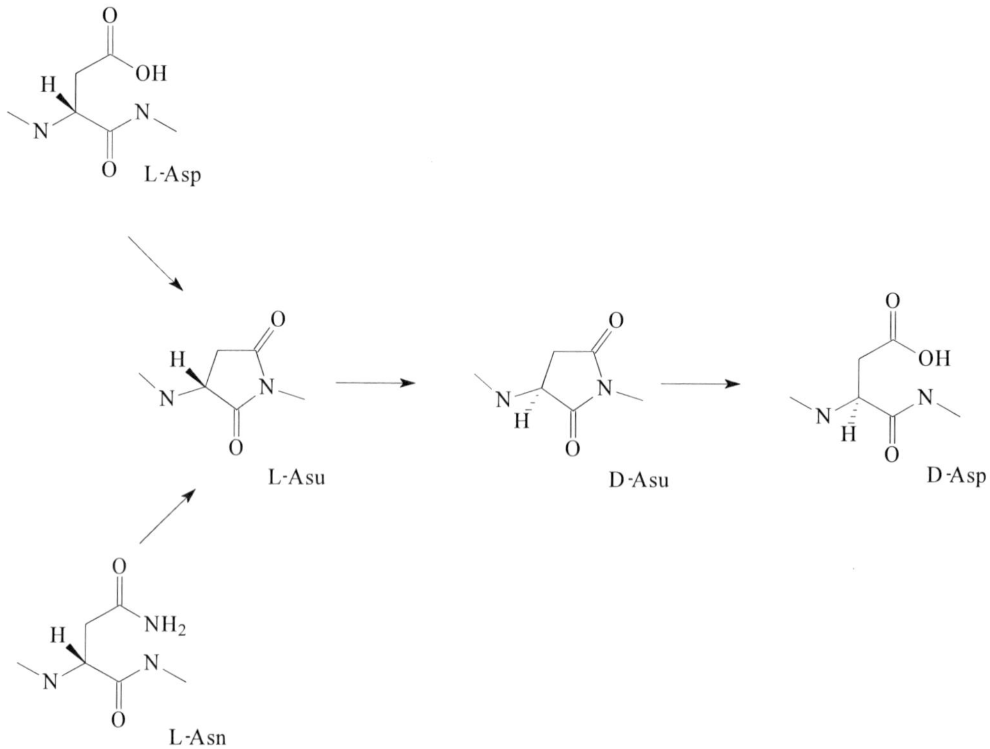

Figure 2. Racemization of L-aspartic acid (L-Asp) and deamination of L-asparagine (L-Asn) via a succinimide intermediate (Asu) to form D-aspartic acid (D-Asp).

2.2. Bound Amino Acids Racemization

Bound amino acids racemizes faster than free ones [25,17]. The simple mechanism via planar carbanion doesn't explain the faster racemization of bound aspartic acid compared with that of serine and threonine [26]. Geiger and Clark demonstrated that in short peptides both aspartic acid (Asp) and asparagine (Asn) racemize to D-Aspartic acid through the formation of a cyclic succinimmide intermediate, named amino succinyl residue (Asu) [27] (Figure 2). Goodfriend noted that the signal measured as Asp is actually the combined responses of Asp + Asn, the latter decomposing to Asp upon acid hydrolysis [28]; to avoid confusion the combined signal is hereafter referred to as Asx. This mechanism is now generally accepted as the major pathway for the racemization of aspartic acid in polypeptides and proteins [26,29].

Molecular structure is very important in determining racemization rate of bound amino acids: for example, bound asparagine racemizes faster than bound aspartic acid, but the reaction rate is strongly influenced by the neighbouring amino acids residues [27,30-31]. Moreover, racemization of amino acids in small peptides is faster than in proteins, where the constraints of the tertiary structure can limit the formation of Asu [32]. Bound aspartic acid racemization follows a very atypical kinetics, having both a fast and a slow component. Some authors suggest that the pattern of Asx racemization may be due to differences in the rate of succinimmide formation from Asp and Asn [28,33].

Among the 20 natural amino acids, the most utilized one is aspartic acid: its rapid kinetics of racemization, before being associated to diagenesis of ancient DNA, has been widely used since 1973 as a relative dating method. What allowed to use the aspartic acid racemization to estimate DNA preservation is the correspondence between the rate of this reaction and the rate of depurination, which is the major responsible of nucleic acids degradation [34-36]. Poinar et al. demonstrated that no DNA sequences could be extracted from samples in which the D/L Asp ratio is higher of 0,08 [8].

Furthermore, because the racemization of Asp is faster than that of other amino acids [36-39], such as alanine (Ala) and leucine (Leu), the extent of racemization of Asp should be the greatest if the mentioned amino acids are of the same age. In contrast, a D/L ratio for Asp which is lower than the D/L ratio for Ala or Leu indicates contamination by exogenous DNA [8,11].

3. Sample Preparation

The classical methods of preparing bones for racemate ratio measurement involve at least three steps: the cleaning of the surface of the sample, the removal of the biomineral matrix and the hydrolysis of the protein into individual amino acids for analysis [40-42].

The cleaning of the surface of the sample is crucial to eliminate possible contaminating proteins (and DNA). The sources of contamination on fossil remains are basically two: 1) the contact with the bone at any point between the excavation and the protein hydrolysis (or DNA extraction); 2) the use of contaminated reagents, labware and lab disposable supplies. To avoid the second source of contamination is important to work in dedicated laboratory facilities, wearing protective clothing and cleaning regularly the work space with oxidants such as bleach and irradiating it with UV lights.

A variety of methods are utilized to remove or exclude contamination from a bone sample: 1) manually cleaning of the surface, 2) physically removing the surface, 3) extracting material only from the interior, 4) washing the surface with acids, 5) irradiating the sample with UV light, 6) exposing the sample to highly concentrated ethanol, 7) exposing the sample to a sodium hypochlorite solution or 8) combinations of these techniques [43].

After decontamination step, proteins of the sample have to be completely hydrolyzed to yield free amino acids. Hydrolysis is usually performed by heating the sample in the presence of high concentration of acids, using thermal or microwave radiation energy, thus obtaining at the same time the bone matrix dissolution. Conventional liquid- or gas-phase hydrolysis is performed in constant boiling HCl in vacuo at 110°C for 24 hours [44,45]. The hydrolysis time can be significantly reduced by using microwave radiation [46-50]. Treatment of the proteins with high concentration of acids at elevated temperatures results in a partial racemization of certain residues. Comparative studies [51,52] demonstrated that racemization induced by hydrolysis through heating or microwave is comparable. Poinar determined that Asp racemization caused by conventional 6N HCl hydrolysis procedure is about 5% [8]. Desalting of hydrolyzate before analysis has been carried out with various methods, such as additioning of hydrofluoric acid to precipitate calcium [28], ultrafiltration [41], the passage through an ion exchange cartridge [52,53], or simply neutralising with NaOH and filtering [52].

4. Amino Acids Analysis

Since most amino acids are not readily detectable, owing to the absence of a strong chromophore and fluorophore, chemical derivatization becomes a necessary procedure to transform the analytes into molecules that can be more easily isolated, separated and detected.

Derivatization is the use of chemical reagents or reactions in order to convert the original structure of the analyte into another molecule or mixture of reaction products.

Different methods were devoted to the determination of amino acids including gas chromatography (GC) [54-63], high-performance liquid chromatography (HPLC) [64-91] with UV [65,78], fluorescence [52,74,78,80,81] or mass spectrometry [83,85,87-90] detection, and capillary electrophoresis (CE) [92-94].

Capillary gas chromatography is ideal for the analysis of the enantiomers because of its inherent high resolution, short analysis time and coupling capabilities with mass spectrometry. In order to be separated by gas chromatography, the amino acids need to be derivatized to become volatile [58-61]. If this derivatization is performed using a chiral reagent, the amino acid enantiomers are converted into diastereoisomers, which can be separated on a non-chiral column. An acylation/esterification procedure using alkyl chloroformates [57] to form diastereomeric N (O,S) ethoxycarbonyl 2-chloropropyl ester derivatives is the most convenient method of derivatization, offering a minimal handling of the sample, the use of an aqueous medium for the reaction, and inexpensive reagents. Engel and Hare [62] described many aspects of the separation of amino acids by GC and concluded that the major breakthrough in the use of this technique to separate enantiomers was the development of optically active stationary phases. Frank at al. developed a chiral column, which was commercialized as Chirasil-Val, and were able to resolve the enantiomers of twenty amino acids [63].

Two main strategies have evolved for the separation of amino acids through High Pressure Liquid Chromatography: an indirect method, based on the reaction of amino acids with a chiral derivatizing agent (CDA) and subsequent separation of the diastereomeric derivatives obtained on a common octadecylsilane phase [52,64,66,75-78,82]; and a direct method, based on the separation of enantiomers by chiral chromatography followed by post-column derivatization to improve their detection [65].

The derivatization of enantiomers to form diastereomeric derivatives separable on achiral columns is the first widely used method in the analysis of chiral drugs and related materials, especially in HPLC. The reasons are the large number of commercially available CDAs and the well-established reactions leading to diastereomeric derivatives with excellent separation and detection possibilities on inexpensive, achiral columns. The reaction should be comparatively fast, but it is essential that the chiral derivatization reaction should proceed quantitatively for both enantiomers, and that racemization should not occur.

The indirect method is very sensitive, so it is suitable for the trace analysis of amino acid enantiomers in complex matrices such as biological samples because of the introduction of a highly sensitive UV–vis or fluorescence tag with the CDA.

In the direct method the enantiomers are separated by the interaction with a chiral stationary phase (CSP) or a chiral selector added to the mobile phase on the stationary phase. Coordination to transition metals (ligand exchange), charge transfer interaction, ion exchange, and inclusion phenomena (host-guest complex) are the most important binding types present in enantioselective adsorption process.

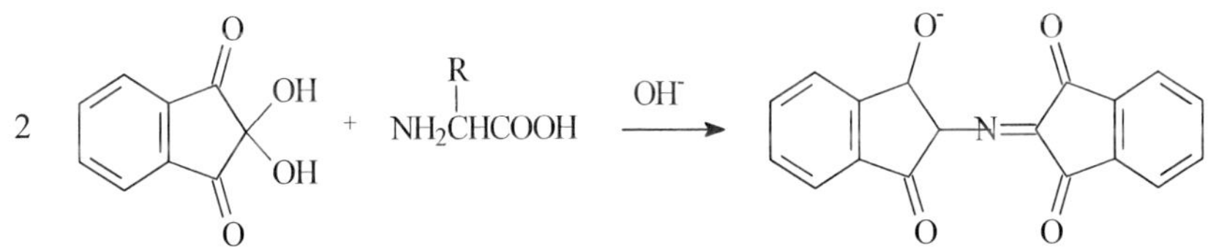

Figure 3. Post-column derivatization with ninhydrin.

The most popular reagent for post-column detection of amino acids is ninhydrin (2,2-Dihydroxyindane-1,3-dione), an oxidating agent which leads to the oxidative deamination of amino groups, giving a complex known as Ruhemann's Purple (Figure 3). In particular α-amino acids yield a purple substance that absorbs maximally at 570 nm, while imino acids, such as proline, yield a yellow product (absorption maximum at 440 nm).

Advantages of the direct method include the simple preparation and the absence of racemization of analytes, but chiral columns are very expensive and above all not universally applicable; furthermore the theoretical plate number of CPS is quite small, and CPS is very sensitive to chromatographic conditions, requiring rather difficult and lengthy method development.

Post-column derivatization reagent must be stable, the yield and signal-to-noise ratio for a given sample remaining constant for at least 8 hours, and the analytical separation must be complete. Furthermore, because the reagent is present in excess relative to the analyte, the analyte's signal could be obliterated by the reagent's strong background signal. All species must remain in solution, including the combined components of the eluents and the reagent(s), as well as the newly formed derivative(s).

4.1. Enantiomeric Derivatization of Amino Acids

The ideal reagent for pre-column derivatization of amino acids should rapidly and completely react with both primary and secondary amino acids in mild conditions, the reagent itself and/or its degradation products should not interfere with the chromatographic separation and the derivatization reaction should not be disturbed by sample matrix components.

The successful determination of optical isomers of amino acids by reversed-phase HPLC is attained with the correct selection of the chiral derivatizing agent and the optimization of chromatographic conditions. Previously it is necessary to study the effect of the nature and the concentration of the organic modifier in the eluent, its composition, and pH on the efficiency and selectivity of the separation.

The major types of reactions for chiral amines, involving amino acids, are based on the formation of amides, carbamates, ureas and thioureas. The reactions with acyl chlorides, chloroformate reagents and *N*-succinimidyl ester proceed rapidly to furnish the corresponding amides and carbamates. Chiral isocyanates and mainly isothiocyanates are good labels with which to produce stable ureas and thioureas. The acid halides are also good labels, because the reactions proceed under mild conditions. However, since the reagent readily undergoes hydrolysis with water in the sample solution, contamination with water in the medium should be avoided. *o*-phthalaldehydes (OPAs) with chiral thiols give highly fluorescent adducts with the primary amino groups of amino acids, ensuring a very low detection limit in the analysis of enantiomers.

4.1.1. Derivatization with Chloroformates

In general, the chloroformates developed for the chiral derivatization of amino groups are fluorescent, and in most cases ensure a very low detection level for amino compounds. The most important chloroformate possessing a reactive functional group is 1-(9-fluorenyl)ethyl chloroformate (FLEC) developed by Einarsson et al. [66]. The reaction conditions (at room temperature in basic solution) are mild, and the resulting fluorescent diastereomers are stable. New (+)- and (−)-4-(6-methoxy-2-naphthyl)-2-butyl chloroformates (NAB-C) were prepared from nabumetone, a prochiral nonsteroidal anti-inflammatory agent, in order to develop an easily detectable chloroformate reagent for the enantiospecific HPLC analysis of amino compounds on achiral stationary phases [95]. The reagent was compared with other reagents containing a 6-methoxy-2-naphthol moiety and was found to be suitable for the analysis of nanogram amounts. New anthracene-based fluorescent reagents, (+)- and (−)-1-(9-anthryl)-2-propyl chloroformates (APOC), were introduced by Thorsen et al. [96] and used for the chiral separation of 17 amino acid enantiomers with a detection limit in the femtomole range.

4.1.2. Derivatization with Isocyanates and Isothiocyanates

Phenylisothiocyanate (PITC) reacts with amino acids to form phenylthiocarbamyl (PTC) derivatives which can be separated on a reversed-phase column and detected on a UV-vis detector.

Isocyanate (–NCO)-bearing reagents generally require more severe reaction conditions than isothiocyanate for completion of the derivatization, *i.e.* a longer reaction time and a higher temperature.

For amino compounds, among the chemically most selective CDAs are isothiocyanates such as 2,3,4,6-tetra-*O*-acetyl-β-D-glucopyranosil isothiocyanate (GITC) [97], 2,3,4,6-tetra-*O*-benzoyl-β-D-glucopyranosyl isothiocyanate (BGIT) and 2,3,4-tri-*O*-acetyl-α-D-arabinopyranosyl isithiocyanate (AITC). Under the reaction conditions applied for the derivatization, isothiocyanates form the corresponding thioureic derivatives with the primary and secondary amino groups of amino acids.

4.1.3. Derivatization with N-haloarylamino Acid Derivatives

Marfey's reagent, the 1-fluoro-2,4-dinitrophenyl-5-L-alanine amide (FDNP-Ala-NH_2 or FDAA), is one of the most widely used reagents for the resolution of amino acid enantiomers. The nucleophilic attack of the α- or β-amino group of the amino acids on the C–F bond activated by the two nitro groups on the aromatic ring results in a smooth reaction to yield diastereomeric aniline derivatives with good UV detectability [98].

Marfey [99] developed a linear gradient of acetonitrile/triethylammonium phosphate (TEAP) buffer (50mM, pH 3.0) for the elution of the five amino acids object of his study. L-diastereomers were eluted from the reversed-phase column before D-diastereomers and all the ten diastereomers had a very similar absorption spectra characterized by a λ_{max} at 340 nm.

The neutral amino acids displayed good resolution between the L and D diastereomers, whereas the separation for hydroxy and acidic amino acid derivatives was very poor. Basic amino acids gave three derivatives and indicated that an α-amino group is essential for the resolution of both diastereomers. For standard α-amino acids, except for a few basic amino acids, the Marfey's method proved to have wide applicability.

In general, FDAA was much favorable as a derivatizing agent than GITC [100], giving better peak shapes and higher resolution.

4.1.4. OPA+Chiral Thiol Reagents

o-phthalaldehyde (OPA) reacts with primary amines in the presence of chiral thiols (figure 4) to form fluorometrically highly active isoindole derivatives.

The reaction with secondary amines is however allowed by the oxidation with sodium hypochlorite. The procedure employs a strongly acidic cation-exchange column for separation of free amino acids followed by post-column oxidation with sodium hypochlorite and derivatization with OPA and thiol compound such as N-acetyl-L-cysteine and 2-mercaptoethanol. The derivatization of primary amino acids is not noticeably affected by the supply of sodium hypochlorite.

From the pioneering work of Roth [101] the reaction of amino acids with *o*-phthaldialdehyde (OPA), in the presence of various SH-group containing agents [70-76], has gained wide acceptance in the chromatographic analysis of amino acids. Because of the instability of the isoindoles obtained from amino acids with the primarily introduced [101] OPA/2-mercaptoethanol (MCE) reagent, as SH- group containing alternatives, 3-mercaptopropionic acid (MPA) [70-72] and N-alkyl-L-cysteines [73-76] were proposed. The advantage of the OPA/MPA and OPA/N-acetyl-L-cysteine (NAC) reagents is that they provide more stable isoindoles, compared to those formed with OPA/MCE, hence in the optical resolution of enantiomeric amino acids OPA/NAC reagents have been extensively utilized.

The OPA + chiral thiol system is widely applied for determination of the amino acid enantiomers contents in living systems, food analyses and geological systems, with fluorescence detection [52,77-80]. The chiral recognition characteristic of the OPA–*N*-alkyl cysteine reagents [75] made it possible to determine in a simple and inexpensive way the L and D amino acids, in extremely different concentrations, in the presence of each other, simultaneously. An OPA and *N*-acetyl-L-Cys derivatization system was applied for the enantiomeric determination of Asp and Ala in fossil samples [52], and for the determination of the D enantiomers of Ser, Thr, Ala, Tyr, Val, Trp and Leu in milk and oyster samples [81]. A novel chiral thiol reagent, *N*-(*tert*-butylthiocarbamoyl)-L-cysteine ethyl ester (BTCC), was developed by Nimura et al. [82] and applied together with OPA for determination of enantiomers of the amino acids Asp, Glu, Ser, Ala and Phe in the aspartate racemase assay. Fluorescence detection was carried out with excitation at 335 nm and emission at 420 nm.

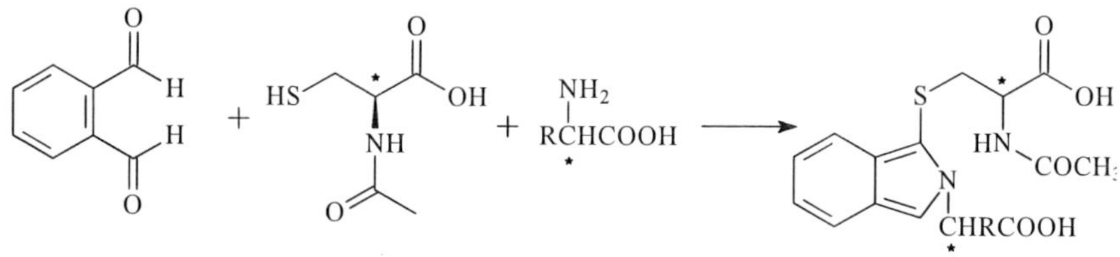

Figure 4. The OPA-NAC derivatization of amino acids.

The successful determination of optical isomers of amino acids by reversed-phase HPLC is attained with the correct selection of the chiral derivatizing agent and the optimization of chromatography conditions. Chernobrovkin et al. [102] studied all the factors that influence the efficiency and selectivity of the separation of amino acid diastereomers: the nature and concentration of the organic modifier in the eluent, its composition, and buffer pH. From their studies emerged that methanol provides a higher selectivity of the separation of amino acid diastereoisomers than acetonitrile. It is expected that, solvating OPA/NAC derivatives of amino acids, methanol forms hydrogen bonds with polar groups of adsorbates and changes the conformation of diastereoisomers, improving their separation. The pH of the mobile phase was varied in the range of 5.0–7.0, because in acidic solutions the OPA derivatives of amino acids are unstable and at pH > 7 a negative effect on the silicate matrix of the sorbent is possible. It is also seen that with increasing pH the retention of the OPA/NAC derivatives of various amino acids decreases insignificantly, and in particular for OPA/NAC-aspartic acid, the retention drastically decreases with increasing pH of the eluent in the range 5.0–6.0. This behaviour of diastereoisomers can be attributed to the presence of the carboxyl group, whose protonation upon decreasing the pH increases the hydrophobicity of adsorbates and their retention. The presence of the additional carboxyl group in the molecule of aspartic acid enhances this effect. An increase in the hydrophobicity of OPA/NAC aspartic acid also leads to a decrease in the efficiency of separation. The retention of the OPA/NAC derivatives of all amino acids increases insignificantly with the concentration of the buffer solution. An increase in the retention times of polar compounds on the addition of inorganic salts to the mobile phase is typical for reversed-phase chromatography and is explained by the salting-out effect. The obtained results indicate that the concentration of the phosphate buffer solution has nearly no effect on the resolution of OPA/NAC derivatives of amino acids [102].

4.1.5. Enzymatic Modification of Amino Acids

Oguri et al. [103] developed a new strategy for the selective determination of D-amino acids, which were enzymatically modified by D-amino acid oxidase. The α-keto acids formed were derivatized with *o*-phenylenediamine in the presence of 2-mercaptoethanol to give fluorescent quinoxalinol derivatives. The D enantiomers of Ala, Leu, Met, Phe and Val in *Ruditapes philippinarum* were analyzed after enzymatic digestion with D-amino acid oxidase and fluorescence detection (excitation at 341 nm, and emission at 413 nm).

5. Applications

Schwarz et al. [104] developed a RP-HPLC method with fluorescence detection in order to facilitate the identification of samples with adequate DNA preservation [8,105] under the assumption that these parameters correlate with degradation processes to which DNA is subjected during diagenesis: using reversed-phase high performance liquid chromatography (RP–HPLC) after pre-column derivatization with o-phthaldialdehyde-N-acetyl-L-cysteine (OPA-NAC), the authors determined abundances of Asp, Glu, Ser, Gly, Ala and Val. Fluorescent detection was monitored at an excitation wavelength of 340 nm and an emission wavelength of 450 nm. The level of amino-acid racemization of aspartic acid (D/L Asp) for all samples was below 0.08, deeming them all to be potentially positive for DNA retrieval.

The degree of bone preservation of five small-size equine skeletons found in the "Casti Amanti" house in Pompeii [106] was documented by measuring the amino acid racemization level. In this case, the authors utilized a modified hydrolysis method [107] based on the activity of a peptidil-D-amino acid hydrolase, enzyme able to free all aspartic residues.

Poinar et al. showed that the extent of racemization of aspartic acid, alanine, and leucine provides criteria for assessing whether ancient tissue samples contain endogenous DNA. They screened a sample set of 26 fossil remains, some of which had yielded endogenous DNA and some others which hadn't. They found that in all cases were D/L Asp ratio was lower than 0,1 but greater than D/L Ala or D/L Leu, DNA was present, while when the level of D/L Asp racemization was higher than 0,15 or lower than D/L ALA or D/L Leu, no endogenous DNA was retrievable. [8].

Marota et al. checked 13 specimens from Egyptian Archeological sites for the presence of residual DNA. They found no amplifiable DNA in all tested samples, even in the less ancient (1400 years old) ones. For all them they determined even the Aspartic acid racemization and found values within 0,09 and 0,26 [108].

Serre et al analyzed the amino acid preservation of 24 Neanderthal and 40 early modern human fossils. They found that endogenous DNA can be amplified from Pleistocene remains when the amino acid content is more than 30,000 parts per million (ppm), the ratio of glycine to aspartic acid between two and ten, and the D/L Asp ratio less than 0.10. Several important Neanderthal fossils, as well as important modern human fossils proved to be too poorly preserved to be likely to allow DNA retrieval. However, four Neanderthal and five early modern human fossils fulfilled the above criteria for amino acid preservation and were thus expected to contain endogenous DNA [10].

Determination of Aspartic acid racemization has became a quite routine analysis in the study of ancient DNA, so that several are the other works in which this tool has been utilized [9,109-113], even if it's usefulness has been questioned by Collins and coauthors [32]. They studied the racemization kinetics of bound aspartic acid and concluded that the D/L ratio in bones depends on the ratio of non helical to triple helical collagen, so the D/L ratio reflects the rate of loss of helix and leaching of soluble material. So they concluded that Asx racemization isn't a reliable proxy for the extent of DNA depurination in bones. In a further study, Collins et al. they analyzed 91 bone and teeth samples from temperate and high-latitude sites and found no correlation between the extent of AAR and DNA amplification success [114].

Conclusion

Although the close correlation between aspartic acid racemization and DNA depurination rates have been recently questioned, the work published till today are not able to discredit racemization data. Moreover, the great sensibility of HPLC with fluorescent detection methods, which can identify amino acids at femtogram levels, make ca. 5 milligrams for sample sufficient for a successful screening, thus without doing more damage to the specimens than is necessary. Finally, the development of faster and automatable methods allows screening large museum collection in a short time [52]. For all the reasons, Aspartic Acid racemization can be still considered a good screening method for ancient DNA preservation.

References

[1] Raup, D. Extinction: bad genes or bad luck? New York, NY: W.W. Norton and Co.; 1991.

[2] Paabo, S. Ancient DNA, extraction, characterization, molecular cloning and enzymatic amplification. *Proceeding of the National Academy of Science USA*, 1989, 86, 1939-1943.

[3] Saiki, RK; Gelfand, DH; Stoffel, S; Scharf, SJ; Higuchi, R; Horn, GT; Mullis, KB; Erlich HA. Primer-directed enzymatic amplification of DNA with a thermostable DNA polymerase. *Science*, 1988, 239, 487-491.

[4] Saiki, RK; Scharf, SJ; Faloona, F; Mullis, KB; Horn, GT; Erlich, HA; Arnheim, N. Enzymatic amplification of β-globin genomic sequences and restriction site analysis for diagnosis of sickle cell anemia. *Science*, 1985, 230, 1350-1354.

[5] Englinton, G; Logan, GA. Molecular preservation. *Philosophical Transactions of the Royal Society of London Series B,* 1991, 333, 315-317.

[6] Hofreiter, M; Serre, D; Poinar, HN; Kuch, M; Pääbo S. Ancient DNA. Nature Reviews/Genetics, 2001, 2, 353-359.

[7] Pääbo, S; Poinar, H; Serre, D; Jaenicke-Despres, V; Hebler, J; Rohland, N; Kuch, M; Krause, J; Vigilant, L; Hofreiter, M. Genetic Analyses from Ancient DNA. *Annual Reviews of Genetics,* 2004, 38, 645-79.

[8] Poinar, HN; Hoss, M; Bada, JL; Paabo, S. Amino Acid Racemization and the preservation of Ancient DNA. *Science*, 1996, 272, 864-866.

[9] Krings, M; Stone, A; Schmitz, RW; Krainitzki, H; Stokening, M; Pääbo S. Neanderthal DNA sequences and the origin of modern humans. *Cell*, 1997, 90, 19-30.

[10] Serre, D; Langaney A; Chech ; Teschler-Nicola M; Paunovic M; Pennecier P; Hofreiter M; Possnert G; Pääbo S. No Evidence of Neanderthal mtDNA Contribution to Early Modern Humans. *Plos Biology*, 2004, 2, 313-317.

[11] Poinar, HN. The genetic Secrets Some Fossil Hold. *Accounts of Chemical Research,* 2002, 35, 676-684.

[12] Rutter, NW; Blackwell, B. Amino acid racemization dating. In N.W. Rutter and NR Catto (eds), Dating Methods for Quaternary Deposits. St. Johns, Newfoundland:

Geological Association of Canada, Memorial University of Newfoundland, 1995, pp 125-164.

[13] Bada, JL. The dating of fossil bones using the racemization of isoleucine. *Earth and Planetary Science Letters,* 1972, 15, 223-231.

[14] Hare, PE; Abelson, PH. Racemization of amino acids in fossil shells. Carnegie Institution of Washington Year Book 66, 1967, 526-528.

[15] Hare, PE; Mitterer, RM. Laboratory Simulation of amino-acid diagenesis in fossils. Carnegie Institution of Washington Year Book 67, 1969, 205-208.

[16] Bada, JL. Racemization of amino acids in nature. *Interdisciplinary Sciences Reviews*, 1982, 7(1), 30-46.

[17] Smith, GG; Evans, RC. The effect of structure and conditions on the rate of racemization of free and bound amino acids. In PE Hare, TC Hoering, and J. King, K (eds). Biogeochemistry of Amino Acids. New York: John Wiley and Sons; 1980; pp. 157-282.

[18] Ebbers, EJ; Ariaans, GJA; Houbiers, JMP; Bruggink, A; Zwanenburg, B. Controlled racemization of optically active organic compounds: Prospects for asymmetric transformation. *Tetrahedron*, 1997, 53, 9417-9476.

[19] Bada, JL. Kinetics of racemization of amino acids as a function of pH. *Journal of the American chemical society*, 1972, 94(4), 1371-1373.

[20] Bada, JL; Shou, MY. Kinetics and mechanism of amino acids racemization in aqueous solution and in bones. In P.E. Hare, TC Hoering, and J. King, K (eds), Biogeochemistry of Amino Acids. New York: John Wiley and Sons; 1980; pp. 235-255.

[21] Pilling, MJ; Seakins, PW. Reaction kinetics. Oxford: Oxford University Press. 1995.

[22] Bada, JL; Schroeder, RA. Amino acids racemization reactions and their geochemical implication. *Die Naturwissenschaften*, 1975, 62, 71-79.

[23] Schroeder, RA; Bada, JL. A review of the geochemical applications of the amino acid racemization reaction. *Earth Science Reviews*, 1976, 12, 347-392.

[24] Smith, GG; Williams, KM; Wonnacott, DM. Factors affecting the rate of racemization of amino acids and their significance to geochronology. *The Journal of Organic Chemistry*, 1978, 43(1), 1-5.

[25] Liardon, R; Ledermann, S. Racemization kinetics of free and protein-bound amino acids under moderate alkaline treatment. *Journal of Agricultural and Food Chemistry,* 1986, 34, 557-565.

[26] Radkiewicz, JL; Zipse, H; Clarke, S; Houk, KN. Accelerated racemization of aspartic acid and asparagines residues via succinimide intermediates: An ab initio theoretical exploration of mechanism. *Journal of the American Chemical Society*, 1996, 118, 9148-9155.

[27] Geiger, T; Clarke, S. Deamidation, isomerization, and racemization at asparaginyl and aspartyl residues in peptides: Succinimide-linked reactions that contribute to protein degradation. *The Journal of Biological Chemistry*, 1987, 262, 785-794.

[28] Goodfriend, GA. Patterns of racemization and epimerization of amino acids in land snail shells over the course of the Holocene. *Geochimica et Cosmochimica Acta*, 1991, 55, 293-302.

[29] van Duin, ACT; Collins, MJ. The effects of conformational constraints on aspartic acid racemization. *Organic Geochemistry*, 1998, 29, 1227-1232.

[30] Brennan, TV; Clarke, S. Spontaneous degradation of polypeptides at aspartyl and asparaginyl residues: effects of the solvent dielectric. *Protein Science*, 1993, 2, 331-338.

[31] Brennan, TV; Clarke, S. Effect of adjacent histidine and cysteine residues on the spontaneous degradation of asparaginyl- and aspartyl-contaminating peptides. *International Journal of Peptide and Protein Research*, 1995, 45, 547-553.

[32] Collins, MJ; Waite, ER; van Duin, ACT. Predicting protein decomposition: the case of aspartic- acid racemization kinetics. *Philosophical Transactions of the Royal Society London B*, 1999, 354(1379), 51-64.

[33] Brinton, KL; Bada, JL. Comment on "Aspartic acid racemization and protein diagenesis in corals over the last 350 years". *Geochimica et Cosmochimica Acta*, 1995, 59(2), 415-416.

[34] Lindahl, T; Nyberg, B. Rate of depurination of native deoxyribonucleic acid. *Biochemistry*, 1972, 11, 3610-3617.

[35] Lindahl, T; Anderson, A. Rate of chain breakage at apurinic sites in double-stranded deoxyribonucleic acid. *Biochemistry*, 1972, 11, 3618-3623.

[36] Bada, JL; Wang, SX; Poinar, HN; Paabo, S; Poinar, GO. Amino acid racemization in amber-entombed insects: implications for DNA preservation. *Geochimica et Cosmochimica Acta,* 1994, 58, 3131-3135.

[37] Bada, JL; Man, EH. Amino acid diagenesis in deep sea drilling project cores: Kinetics and mechanisms of some reactions and their applications in geochronology and in paleotemperature and heat flow determinations. *Earth Science Reviews*. 1980, 16, 21-25.

[38] Bada, JL. Racemization of amino acids. In: G. C. Barrett (Ed.), Chemistry and Biochemistry of Amino Acids, Chapman and Hall, London, 1985, pp 399-414.

[39] Bada, JL. Amino acids racemization dating of fossil bones. *Annual Review of Earth and Planetary Science*, 1985, 13, 241-268.

[40] Bada, JL; Wang, XS; Hamilton, H. Preservation of key biomolecules in the fossil record: current knowledge and future challenges. Philosophical Transaction of the Royal Society London B, 1999, 354, 77-78.

[41] Kimber, RW; Hare, PE. Wide range of racemization of amino acids in peptides from human fossil bone and its implications for amino acid racemization dating. *Geochimica et Cosmochimica Acta*, 1992, 56, 739-743.

[42] Van Klinken, GJ; Mook, WG. Preparative high-performance liquid chromatographic separation of individual amino acids derived from fossil bone collagen. *Radiocarbon,* 1990, 32, 155-164.

[43] Kemp, BM; Smith, DG. Use of bleach to eliminate contaminating DNA from the surface of bones and teeth. *Forensic Science International*, 2005, 154, 53-61.

[44] Spackman, DH; Stein, WH; Moore, S. Automatic Recording Apparatus for Use in the Chromatography of Amino Acids. *Analytical Chemistry*, 1958, 30, 1190-1206.

[45] Hirs, CH; Stein, WH; Moore, S. The amino acid composition of ribonuclease. *Journal of Biological Chemistry*, 1954, 211, 941-950.

[46] Fischer, LB. Microwave Dissolution of Geologic Material: Application to isotope dilution analysis. *Analytical Chemistry*, 1986, 58, 261-263.

[47] Pecavar, A; Prosek, M; Fercejtemeljotov, D; Marsel, J. Quantitative evaluation of amino acids using microwave accelerated hydrolysis *Chromatographia*, 1990, 30, 159-162.

[48] Lahm, H-W; Lergier, W; Manneberg, M; Knorr, R. An Improved Method for Fast Amino-Acid Analysis. *Journal of Protein Chemistry*, 1988, 7, 258-259.

[49] Peter, A; Laus, G; Tourwe, D; Gerlo, E; Vanbinst, G. An evaluation of microwave heating for the rapid hydrolysis of *peptide samples* for chiral amino acid analysis. *Peptide Research*, 1993, 6, 48-52.

[50] Joergensen, L; Thestrup, HN. *Determination of amino acids in biomass and protein* samples by microwave hydrolysis and ion-exchange chromatography. *Journal of Chromatography A*, 1995, 706, 421-428.

[51] Stenberg, M; Marko-Varga, G; Oste, R. *Racemization* of *amino acids* during classical and microwave oven hydrolysis - Application to aspartame and Maillard reaction system. *Food Chemistry*, 2001, 74, 217-224.

[52] Iuliani, P, Di Federico, L, Fontecchio, G, Carlucci, G. RP-HPLC method with fluorescence detection for amino acids D/L ratio determination in fossil bones as a marker of DNA preservation. *Journal of Separation Science*, 2010, 33, 2411-2416.

[53] Spanik, I; Horvathova, G; Janacova, A; Krupcik, J. On the use of solid phase exchangers for isolation of amino acids from liquid samples and their enantioselective gas chromatographic analysis. *Journal of Chromatography A*, 2007, 1150, 145-154.

[54] Dauner, M; Sauer, U. GC-MS analysis of amino acids rapidly provides rich information for isotopomer balancing. *Biotechnology Progress*, 2008, 16, 642-649.

[55] Deng, C; Yin, X; Zhang, L; Zhang, X. Development of microwave-assisted derivatization followed by gas chromatography/mass spectrometry for fast determination of amino acids in neonatal blood samples. *Rapid Communication in Mass Spectrometry*, 2005, 19, 2227-2234.

[56] Nozal, MJ; Bernal, JL; Toribio, ML; Diego, JC; Ruiz, A. Rapid and sensitive method for determining free amino acids in honey by gas chromatography with flame ionization or mass spectrometric detection. *Journal of Chromatography A*, 2004, 1047, 137-146.

[57] Bertrand, M; Chabin, A; Brack, A; Westall, F. Separation of amino acid enantiomers via chiral derivatization and non-chiral gas chromatography. *Journal of Chromatography A,* 2008, 1180, 131-137.

[58] Duncan, MW; Poljac, A. Amino Acid analysis of peptides and proteins on the femtomole scale by gas chromatography/mass spectrometry. *Analytical Chemistry*, 1998, 70, 890-896.

[59] Mackenzie, SL. Gas chromatographic analysis of amino acids as the N-heptafluorobutyryl isobutyl esters. *Journal of the Association of Official Analytical Chemistry*, 1987, 70, 151-160.

[60] Oh, C-H; Kim, J-H; Kim, K-R; Brownson, DM; Mabry, TJ. Simultaneous gas chromatographic analysis of non-protein and protein amino acids as N(O,S)-isobutyloxycarbonyl tert.-butyldimethylsilyl derivatives. *Journal of Chromatography A,* 1994, 669, 125-137.

[61] Woo, KL; Lee, DS. Capillary gas chromatographic determination of proteins and biological amino acids as N(O)-tert.-butyldimethylsilyl derivatives. *Journal of Chromatography B,* 1995, 665, 15-25.

[62] Engel, MH; Hare, PE. Gas-liquid chromatographic separation of amino acids and their derivatives. In G.C. Barret (Ed), Chemistry and Biochemistry of the Amino Acids. London: Chapman and Hall; 1985; pp 462-479.

[63] Frank, H; Nicholson, GJ; Bayer, E. Rapid gas chromatographic separation of amino acid enantiomers with a novel chiral stationary phase. *Journal of Chromatographic Science,* 1977, 15, 174-176.

[64] Aswad, DV. Determination of D- and L-aspartate in amino acid mixtures by high-performance liquid chromatography after derivatization with a chiral adduct of o-phthaldialdehyde. *Analytical Biochemistry*, 1984, 137, 405-409.

[65] Ding, GS; Liu, Y; Cong, RZ; Wang, JD. Chiral separation of enantiomers of amino acid derivatives by high-performance liquid chromatography on a norvancomycin-bonded chiral stationary phase. *Talanta,* 2004, 62, 997-1003.

[66] Einarsson, S; Josefsson, B; Moller, P; Sanchez, D. Separation of amino acid enantiomers and chiral amines using pre-column derivatization with (+)-1-(9-fluorenyl)ethyl chloroformate and reversed-phase liquid chromatography. *Analytical Chemistry*, 1987, 59, 1191–1195.

[67] Péter, A; Olajos, E; Casimir, R; Tourwe, D; Broxterman, QB; Kaptein, B; Armstrong, DW. High-performance liquid chromatographic separation of the enantiomers of unusual alpha-amino acid analogues. *Journal of Chromatography A,* 2000, 871, 105–113.

[68] Péter, A; Lázár, L; Fülöp, F; Armstrong, DW. High-performance liquid chromatographic enantioseparation of beta-amino acids. *Journal of Chromatography A,* 2001, 926, 229–238.

[69] Péter, A; Török, R; Wright, K; Wakselman, M; Mazaleyrat, J-P. Liquid chromatographic enantioseparation of spin-labelled beta-amino acids. *Journal of Chromatography A,* 2003, 1021, 1–10.

[70] Godel, H; Graser, T; Földi, P; Pfaender, P; Fürst, P. Measurement of free amino acids in human biological fluids by high-performance liquid chromatography. *Journal of Chromatography,* 1984, 297, 49–61.

[71] Palermo, S; de Marchis, M; Prati, M; Fugassa, E. HPLC analysis of free amino acids and amino acids of total proteins in cultured cells: an application to the study of rat Sertoli cell protein metabolism. *Analytical Biochemistry*, 1992, 202, 152–158.

[72] van Eijk, NMH; Rooyakkers, DR; Deutz, NEP. Rapid routine determination of amino acids in plasma by high-performance liquid chromatography with a 2-3 microns Spherisorb ODS II column. *Journal of Chromatography*, 1993, 620, 143–148.

[73] Buck, RH; Krummen, K. High-performance liquid chromatographic determination of enantiomeric amino acids and amino alcohols after derivatization with o-phthaldialdehyde and various chiral mercaptans. Application to peptide hydrolysates. *Journal of Chromatography*, 1987, 387, 255–265.

[74] Fujiwara, M; Ishida, Y; Nimura, N; Toyama, A; Kinoshita, T. Postcolumn fluorometric detection system for liquid chromatographic analysis of amino and imino acids using o-phthalaldehyde/N-acetyl-L-cysteine reagent. *Analytical Biochemistry*, 1987, 166, 72–78.

[75] Zhao, M; Bada, JL. Determination of alpha-dialkylamino acids and their enantiomers in geological samples by high-performance liquid chromatography after derivatization

with a chiral adduct of o-phthaldialdehyde. *Journal of Chromatography A,* 1995, 690, 55–63.

[76] Hashimoto, A; Nishikawa, T; Oka, T; Takahashi, K; Hayashi, T. Determination of free amino acid enantiomers in rat brain and serum by high-performance liquid chromatography after derivatization with N-tert-butyloxycarbonyl-L-cysteine and o-phthaldialdehyde. *Journal of Chromatography*, 1992, 582, 41–48.

[77] Pereira, V; Pontes, M; Câmara, JS; Marques, JC. Simultaneous analysis of free amino acids and biogenic amines in honey and wine samples using in loop orthophthalaldeyde derivatization procedure. *Journal of Chromatography A*, 2008, 1189, 435-443.

[78] Schwarz, EL; Roberts, WL ; Pasquali, M. Analysis of plasma amino acids by HPLC with photodiode array and fluorescence detection. *Clinica Chimica Acta*, 2005, 354, 83-90.

[79] Zacharis, CK; Theodoridis, GA; Voulgaropoulos, AN. Coupling of sequential injection with liquid chromatography for the automated derivatization and on-line determination of amino acids. *Talanta*, 2006, 69, 841-847.

[80] Rigobello-Masini, M; Penteado, JCP; Liria, CW; Miranda, MTM; Masini, JC. Implementing stepwise solvent elution in sequential injection chromatography for fluorimetric determination of intracellular free amino acids in the microalgae Tetraselmis gracilis. *Analytica Chimica Acta,* 2008, 628, 123-132.

[81] Rubio-Barosso, S; Santos-Delgado, MJ; Martin-Olivar, C; Polo-Diez, LM. Indirect chiral HPLC determination and fluorimetric detection of D-amino acids in milk and oyster samples. *Journal of Dairy Science*, 2006, 89, 82–89.

[82] Nimura, N; Fujiwara, T; Watanabe, A; Sekine, M; Furuchi, T; Yohda, M; Yamagishi, A; Oshima, T; Homma, H. A novel chiral thiol reagent for automated precolumn derivatization and high-performance liquid chromatographic enantioseparation of amino acids and its application to the aspartate racemase assay. *Analytical Biochemistry*, 2003, 315, 262–269.

[83] van Eijk, HMH; Rooyakkers, DR; Soeters, PB; Deutz, EP. Determination of amino acid isotope enrichment using liquid chromatography-mass spectrometry. *Analytical Biochemistry,* 1999, 271, 8.

[84] Fekkes, D; van Dalen, A; Edelman, M; Voskuilen, A. Validation of the determination of amino acids in plasma by high-performance liquid chromatography using automated pre-column derivatization with o-phthaldialdehyde. *Journal of Chromatography B,* 1995, 669, 177.

[85] Shimbo, K; Oonuki, T; Yahashi, A; Hirayama, K; Miyano, H. Precolumn derivatization reagents for high-speed analysis of amines and amino acids in biological fluid using liquid chromatography/electrospray ionization tandem mass spectrometry. *Rapid Communication in Mass Spectrometry*, 2009, 23, 1483-1492.

[86] López-Cervantes, J; Sánchez-Machado, DI; Rosas-Rodrıgues, JA. Analysis of free amino acids in fermented shrimp waste by high-performance liquid chromatography. *Journal of Chromatography A,* 2006, 1105, 106-110.

[87] Ozcan, S; Senyuva, H. Improved and simplified liquid chromatography/atmospheric pressure chemical ionization mass spectrometry method for the analysis of underivatized free amino acids in various foods. *Journal of Chromatography A,* 2006, 1135, 179-185.

[88] Petritis, K; Chaimbault, P; Elfakir, C; Dreux, M. Parameter optimization for the analysis of underivatized protein amino acids by liquid chromatography and ionspray tandem mass spectrometry. *Journal of Chromatography A*, 2000, 896, 253-263.
[89] Qu, J; Wang, Y; Luo, G; Wu, Z; Yang, FC. Validated quantitation of underivatized amino acids in human blood samples by volatile ion-pair reversed-phase liquid chromatography coupled to isotope dilution tandem mass spectrometry. *Analytical Chemistry,* 2002, 74, 2034-2040.
[90] Fonteh, AN; Harrington, RJ; Harrington, MG. Quantification of free amino acids and dipeptides using isotope dilution liquid chromatography and electrospray ionization tandem mass spectrometry. *Amino Acids* 2007, 32, 203-212.
[91] Lozanov, V; Petrov, S; Mitev, V. Simultaneous analysis of amino acid and biogenic polyamines by high-performance liquid chromatography after pre-column derivatization with N-(9-fluorenylmethoxycarbonyloxy)succinimide. *Journal of Chromatography A,* 2004, 1025, 201-208.
[92] Zhang, H; Potier, IL; Smadja, C; Zhang, J; Taverna, M. Fluorescent detection of peptides and amino acids for capillary electrophoresis via on-line derivatization with 4-fluoro-7-nitro-2,1,3-benzoxadiazole. *Analytical and Bioanalytical Chemistry*, 2006, 386, 1387-1394.
[93] Qu, Q; Tang, X; Wang, C; Yang, G; Hu, X; Lu, X; Liu, Y; Li, S; Yan, C. Characterization and application of a new ultraviolet derivatization reagent for amino acids analysis in capillary electrophoresis. *Analytica Chimica Acta*, 2006, 572, 212-218.
[94] Kovács, A; Simon-Sarkadi, L; Ganzler, K. Determination of biogenic amines by capillary electrophoresis. *Journal of Chromatography A*, 1999, 836, 305-313.
[95] Büshges, R; Devant, R; Mutschler, E; Spahn-Langguth, H. 4-(6-methoxy-2-naphthyl)-2-butyl chloroformate enantiomers: new reagents for the enantiospecific analysis of amino compounds in biogenic matrices. *Journal of Pharmaceutical and Biomedical Analysis,* 1996, 15, 201–220.
[96] Thorsen, G; Engström, A; Josefsson, B. Enantiomeric determination of amino compounds with high sensitivity using the chiral reagents (+)- and (-)-1-(9-anthryl)-2-propyl chloroformate. *Journal of Chromatography A*, 1997, 786, 347–354.
[97] Hefnawy, MM; Stewart, JT. HPLC separation of metirosine enantiomers as methyl esters derivatized with 2,3,4,6-tetra-*O*-acetyl-β-D-glucopyranosyl isothiocyanate. *Journal of Liquid Chromatography and Related Technologies*, 1998, 21, 381–389.
[98] Bhushan, R; Brückner, H. Marfey's reagent for chiral amino acid analysis: a review. *Amino acids,* 2004, 27, 231-247
[99] Marfey, P. Determination of D-amino acids. II. Use of a bifunctional reagent, 1,5-difluoro-2,4-dinitrobenzene. *Carlsberg Research Communication*, 1984, 49, 591–596.
[100] Péter, A; Tóth, G; Török, G; Tourwé, D. Separation of enantiomeric beta-methyl amino acids and of beta-methyl amino acid containing peptides. *Journal of Chromatography A,* 1996, 728, 455–465.
[101] Roth, M. Micro-analysis of amino acids. *Analytical Chemistry*, 1971, 43, 880–882.
[102] Chernobrovkin, MG; Anan'eva, IA; Shapovalova, EN; Shpigun, OA. Determination of amino acid enantiomers in pharmaceuticals by reversed-phase high-performance liquid chromatography. *Analytical Chemistry*, 2004, 59, 64-72.

[103] Oguri, S; Nomura, M; Fujita, Y. A new strategy for the selective determination of D-amino acids: enzymatic and chemical modifications for pre-column derivatization. *Journal of Chromatography A*, 2005, 1078, 51–58.

[104] Schwarz,C; Debruyne, R; Kuch, M; McNally, E; Schwarcz, H; Aubrey, AD; Bada, J; Poinar, H. New insights from old bones: DNA preservation and degradation in permafrost preserved mammoth remains. *Nucleic Acids Research*, 2009, 37, 3215–3229

[105] Poinar, HN; Stankiewicz, BA. Protein preservation and DNA retrieval from ancient tissues. *Proceeding of the National Academy of Science USA,* 1999, 96, 8426–8431.

[106] Di Bernardo, G; Galerisi, U; Del Gaudio, S; D'Aniello, A; La nave, C; De Robertis, MT; Cascino, A; Cipollaro, M. Genetic characterization of Pompeii and Herculaneum equidae buried by Vesuvius in 79 AD. *Journal of Cellular Physiology,* 2004, 199, 200–205.

[107] D'Aniello, A; Petrucelli, L; Gardner, C; Fisher, G. Improved method for hydrolyzing proteins and peptides without inducing racemization and for determining their true D-amino acid content. *Analytical Biochemistry*, 1993, 213, 290–295.

[108] Marota, I; Basile, C; Ubaldi, M; Rollo, F. DNA Decay Rate in Papiry and Human Remains From Egyptian Archeological Sites. *American Journal of Physical Anthropology,* 2002, 117, 310-318.

[109] Hyndman, KA; Evans, DH. Endothelin and endothelin converting enzyme *Journal of Experimental Biology,* 2007, 210, 4286-97.

[110] Harbeck, M; Ritz-Timme, S; Schröder, I; Oehmichen, M; von Wurmb-Schwark, N. Degradation of biomolecules: a comparative study of diagenesis of DNA and proteins in human bone tissue. *Anthropol. Anz,* 2004, 62, 387-96.

[111] Pusch, CM; Blin, N; Broghammer, M; Nicholson, GJ; Scholz M. Adaptor-mediated amplification of minute amounts of severely fragmented ancient nucleic acid *Journal of Applied Genetics*, 2000, 41, 303-15.

[112] Pusch, CM; Broghammer, M; Scholz M. Cremation practices and the survival *Anthropol. Anz*, 2000, 58, 237-51.

[113] Bachmann, L; Scholz, M; Broghammer, M; Giddings, I; Pusch, CM. Voltage-induced release of nucleic acid *Electrophoresis*, 2000, 21, 1488-92.

[114] Collins MJ, Penkman KE, Rohland N, Shapiro B, Dobberstein RC, Ritz-Timme S, Hofreiter M. Is amino *Proceeding in Biological Science*, 2009, 276, 2971-7.

In: Tyrosine and Aspartic Acid
Editors: J. E. Jones and D. M. Morano

ISBN: 978-1-62100-752-4

Chapter 5

MOLECULAR DYNAMICS SIMULATION FOR STEREOINVERSION OF ASPARTIC ACID RESIDUES

Akifumi Oda[1,2,*], Kana Kobayashi[1], and Ohgi Takahashi[1]
[1]Faculty of Pharmaceutical Sciences, Tohoku Pharmaceutical University, Japan
[2]Institute for Protein Research, Osaka University, Japan

ABSTRACT

Recently, stereoinversions of aspartic acid residues were observed in some proteins and peptides. Because D-aspartic acids were found in crystallin from cataract patients and in β-amyloid from Alzheimer's dementia patients, it is suspected that the stereoinversion of aspartic acid residues is relevant in age-related diseases. The stereoinversion of aspartic acid residues should induce structural changes in their associated proteins and peptides, and therefore, structural investigations of these modified proteins and peptides are essential. Furthermore, proteins that incorporate D-aspartic acid residues (e.g., crystallin and β-amyloid) tend to aggregate, making experimental investigations of their atomic-level structures difficult. Thus, structural bioinformatics methods play an important role in studies of D-aspartic acid residues. We employed the molecular dynamics (MD) method, which is one of the most useful structural bioinformatics procedures, to investigate the stereoinversion-related properties of proteins. In this chapter, an evaluation of various parameters used in MD simulations was performed. Because parameters were developed for L-aspartic acid, their usefulness for D-aspartic acids was unknown. We computationally constructed peptides including D-aspartic acid residues and performed MD simulations with these peptides. The results suggest that the parameters developed for L-aspartic acid are also appropriate for D-aspartic acid residues.

[*] Correspondence concerning this article should be addressed to Akifumi Oda, E-mail: oda@tohoku-pharm.ac.jp, Fax: +81-22-275-2013.

INTRODUCTION

Aspartic acid residues are known to be susceptible to stereoinversion in human proteins [1-3]. Previously, it was thought that all amino acids in the human body exhibit the L-form. However, recently, D-amino acids have been identified in the human body to exist both independently as free amino acids and as residues in proteins [1]. Investigations into the mechanisms by which D-amino acids are produced have been conducted both experimentally and theoretically, and the results suggest that these residues are mainly formed by nonenzymatic reactions. It has been reported that D-amino acid residues in human proteins increase with aging, and the influence of ultraviolet exposure was suggested as a possible cause for the stereoinversion of amino acid residues leading to such products [4]. Of the possible D-amino acid residues, D-aspartic acid is the one most widely observed in proteins. The proposed mechanism for the stereoinversion of aspartic acid residues is shown in Figure 1. As seen in the figure, stereochemically labile cyclic intermediates are formed, which are in equilibrium with the corresponding L- or D-aspartic acid residues or β-aspartic acid residues. The tendency toward stereoinversion or constitutional isomerization is considered to be determined by the three-dimensional structural environment around the aspartic acid residues. In addition, we clarified the catalytic activities of water molecules that promote formation of the intermediate [2]. D-Amino acid residues are present in many proteins and peptides, such as amyloid β peptides in the brain of Alzheimer's dementia patients [5-7], α-crystallin in the lens of cataract patients [8], and skin proteins in the elderly [4]. Consequently, it is considered that these stereoinversion and isomerization events play important roles in age-related diseases.

The stereoinversion of aspartic acid residues changes the physicochemical properties of the proteins in the vicinity of the inverted residues (Figure 2). The exchange between a bulky side chain and an α hydrogen alters the steric environment, and the displacement of the carboxyl moiety influences the electrostatic environment. These local environmental changes induce structural changes not only around the aspartic acid residue but also throughout the protein. The resulting global changes cause protein aggregation, leading to insoluble aggregates. The formation of fibrils in living cells may also result from the stereoinversion of aspartic acid residues. In general, the mutation of only one residue frequently has a significant consequence for both protein structure and function. For example, the mutation of a single residue affects the enzymatic activity of drug-metabolizing enzymes, leading to individual differences in the efficacy of medicines as well as their adverse effects [9-11]. The stereoinversion of aspartic acid residues also affects protein function similar to what is observed for amino acid mutations even though the resulting isomers are compositionally equivalent. Although the importance of stereoinversion of amino acid residues has been reported, sufficient atomic/molecular level investigations have not been performed. This is probably due to the reduced solubility of denatured proteins in comparison to their native counterparts, which makes experimental structural studies, especially at the atomic level, quite difficult. Therefore, the use of structural bioinformatics provides a practical alternative for studying these proteins. Furthermore, for structurally unknown proteins and peptides, structural bioinformatics studies are frequently useful.

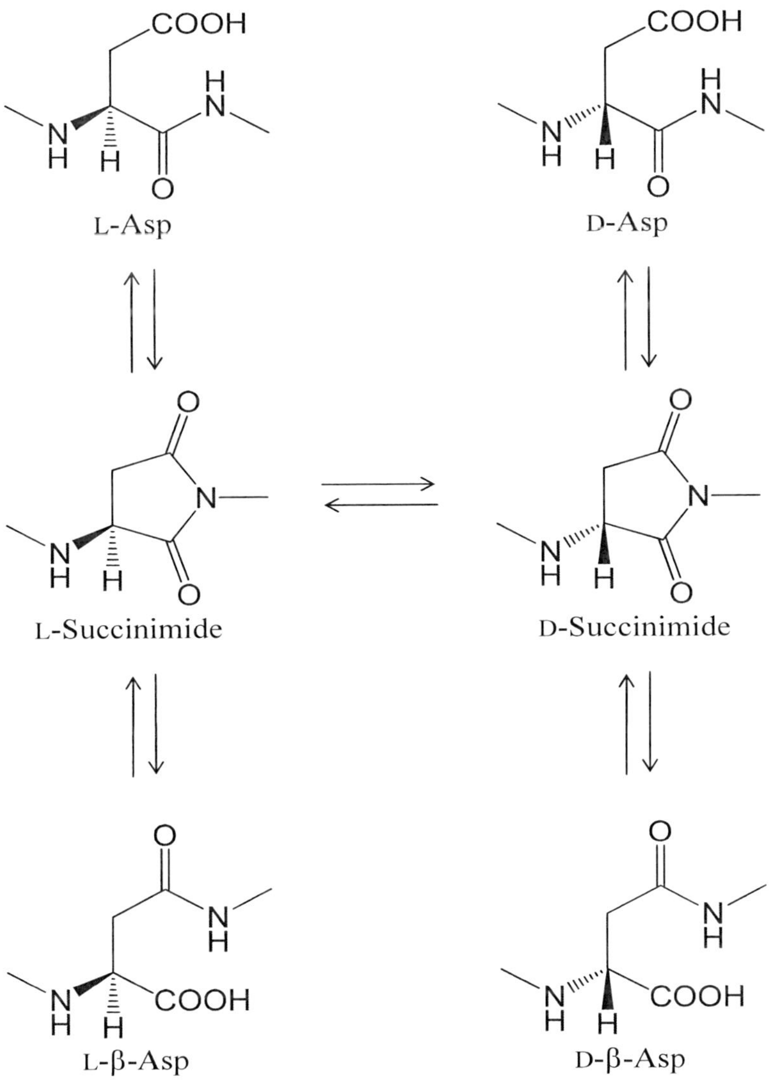

Figure 1. Stereoinversion and isomerization of the aspartic acid residue.

The molecular simulation technique, which is one of the structural bioinformatics methods, is used for the investigation of structural and functional modifications to proteins caused by environmental changes. In a molecular simulation, the behavior of the atomic- or molecular-level structure of a protein is simulated according to physical laws. Molecular dynamics (MD) simulation is one of the most widely used molecular simulation techniques. In MD simulations, the behavior of the molecule can be simulated using the numerical solution of Newton's equation of motion. The outline of an MD simulation is illustrated in Figure 3. As shown in the figure, first, the potential energy of each atom included in the system is calculated. Although the atomic or molecular energies should be calculated by quantum chemical methods, such calculations of biopolymers are difficult because of high computational costs.

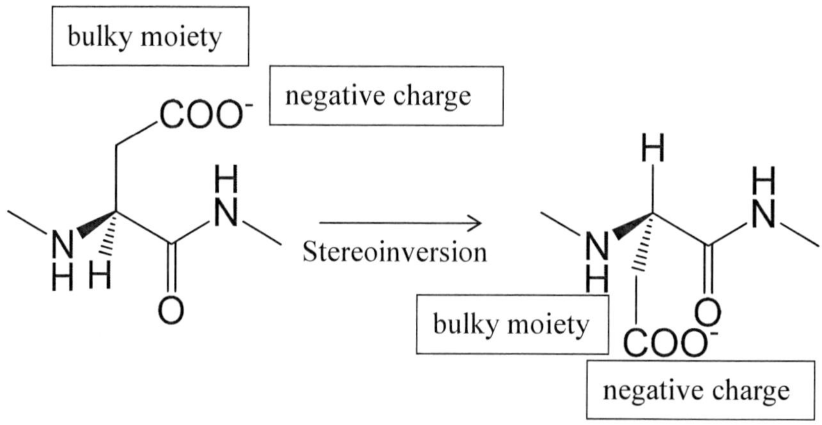

Figure 2. Influence of stereoinversion of the aspartic acid residue on the protein environment.

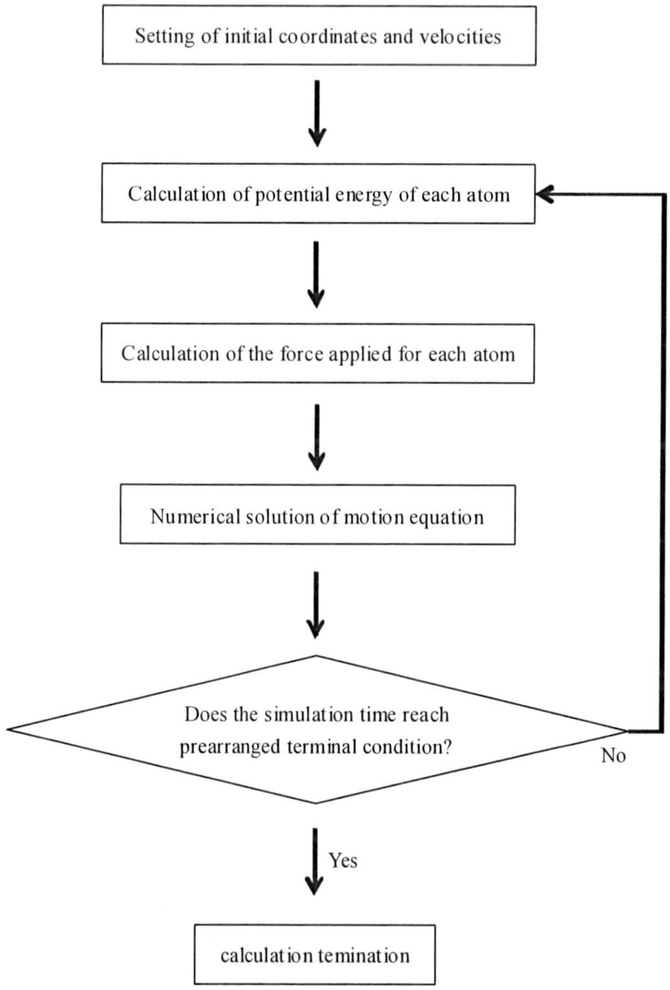

Figure 3. Flow chart of an MD simulation.

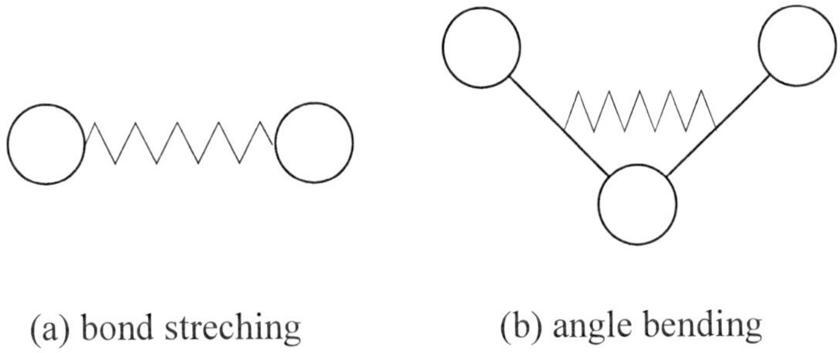

Figure 4. Harmonic-oscillator approximation of bond stretching and angle bending.

Thus, potential energies are calculated using classical force-field approximations. For example, the potential energy E_{total} using the AMBER force field [12-18], which is one of the most widely used force fields for biopolymers, is as follows (the equation is used in ff99[15,16] and ff03[17,18]).

In this equation, the first and second terms, which are related to bond stretching and angle bending, respectively, express harmonic-oscillator approximations (Figure 4). The parameters r_{eq} and θ_{eq} are the equilibrium bond length and bond angle, respectively, and K_r and K_θ are force constants.

$$E_{total} = \sum_{bonds} K_r \left(r - r_{eq}\right)^2 + \sum_{angles} K_\theta \left(\theta - \theta_{eq}\right)^2 + \sum_{dihedrals} \frac{V_n}{2}\left[1 + \cos(n\phi - \gamma)\right] + \sum_{i<j}\left(\frac{A_{ij}}{R_{ij}^{12}} - \frac{B_{ij}}{R_{ij}^{6}} + \frac{q_i q_j}{\varepsilon R_{ij}}\right) \quad (1)$$

The third term is the torsional term, where V_n is the energy barrier of torsional motion, n is the periodicity, and γ is the phase.

The last term describes nonbonding interactions, i.e., van der Waals interactions, and Coulombic interactions. In this equation, r, θ, ϕ, and R_{ij} are the structural features of the system (bond length, bond angle, torsional angle, and distance between atoms i and j, respectively), and q_i and q_j are the atomic charges of atoms i and j, respectively. When the potential energy is obtained by eq. (1), the force **F** can be calculated by differentiating with respect to the position. Using the force, positional vector **x** can be derived from Newton's equation of motion (2). In the MD simulation, eq. (2) is numerically solved.

$$\mathbf{F} = m\frac{\mathrm{d}^2\mathbf{x}}{\mathrm{d}t^2} \quad (2)$$

MD simulations are used for a wide variety of biological studies, such as protein folding predictions [19], protein structure refinements [20], structural analysis of induced fit caused by ligand binding [21], and investigations into the influence of amino acid mutations on protein function [22]. It plays an important role not only in basic biology, but also medical and pharmaceutical sciences. On the other hand, because potential-energy calculations using eq. (1) are indispensable to classical MD simulations, the preparation of force-field parameters is required. For L-amino acids, the form generally found in proteins, force-field parameters are available and have been sufficiently tested. However, parameters for molecules other than standard amino acids and nucleic acids, such as small drug molecules, cofactors, and nonnatural amino acids, are frequently not available. For calculations of these molecules, the determination of new parameters, reuse of existing parameters, or both are required. Previously, our group determined new parameters for the heme moiety [23]. D-Aspartic acid is one of the molecules for which force-field parameters have not been derived. As mentioned above, D-aspartic acid residues are found in several proteins, including those associated with diseases of older adults. Therefore, the development of tools that enable investigations of the three-dimensional structures of such proteins is important. Thus, the preparation of force-field parameters for D-aspartic acid plays an important role in structural bioinformatics studies of age-related diseases.

In this manuscript, the preparation of force-field parameters for D-aspartic acid is discussed. Two approaches are used: one is the reuse of L-aspartic acid parameters for D-aspartic acid, and the other is the use of parameters for small molecules.

2. Force-Field Parameters for D-Aspartic Acid Residue

For classical MD simulations, force-field parameters are indispensable to molecular energy calculations. There are many types of molecular force fields used for this purpose. For example, MM2[24], MMFF[25], OPLS force field [26-27], CHARMm[28], AMBER force field[12-18], and GROMOS[29] are approximation functions developed to calculate molecular energy, and these functions have both advantages and disadvantages. Among these force fields, the AMBER force field is one of the most widely used for calculations of biopolymers. The AMBER force field is used in the molecular mechanics and molecular dynamics program package AMBER [30]. In the AMBER force field, parameters for natural amino acids and nucleic acids were determined and extensively examined. AMBER force-field parameters are used not only in the AMBER program but also in other molecular mechanics and molecular dynamics programs, (e.g., CHARMm and GROMACS) and molecular modeling programs (e.g., SYBYL) because of their high reliability for biopolymers. AMBER force-field parameters for natural amino acids were developed in 1984 [12], and they have been continually improved. The atom type and parameters were enhanced in 1995 [14], and at present, ff99[15], its improvement (ff99SB) [16], and ff03[17,18] are commonly used. In ff99SB and ff03, there are both advantages and disadvantages [31]. In this manuscript, ff99SB is used for the AMBER normal amino acid force field.

On the other hand, for energy calculations of the molecules whose energies cannot be calculated using the AMBER force field for normal amino acids and nucleic acids, such as small drug molecules and cofactors, general AMBER force field (GAFF) was developed [32]. GAFF can be used for every type of molecule, and it is easy to use in combination with the Antechamber tool [33].

GAFF and Antechamber assist structural bioinformatics studies of proteins, including drugs, cofactors, and proteins that include nonnatural amino acids. Because these molecules play important roles in the human body, GAFF and Antechamber are also useful for drug design and development.

For molecular simulations of proteins including D-aspartic acid residues using the AMBER force field, there are two types of parameter settings. One is the application of L-aspartic acid parameters for D-aspartic acid residues, and the other is the application of GAFF. In this section, two examples are described. One involves MD simulations of comparatively long (including 19 residues) peptides in which L-aspartic acid parameters were used for D-aspartic acid residues [34]. The other encompasses MD simulations of hexapeptides, which are enzyme substrates, and in these simulations, GAFF was used [35].

2.1. Amber Amino Acid Parameters

In this section, AMBER ff99SB force-field parameters for L-aspartic acids are used for D-aspartic acid residues. The atom type settings of the AMBER force field for the L-aspartic acid residue are shown in Figure 5. Not only force-field parameters but also atomic charges for the L-aspartic acid residue were used for D-aspartic acid residues.

The structure of the D-aspartic acid residue was constructed from the L-aspartic acid structure using the tleap module of AmberTools [30]. That is, the β carbon and α hydrogen of L-aspartic acid (Figure 6) were exchanged, and the side chain atoms were complimented by tleap. The generated structures can then be used in the AMBER program.

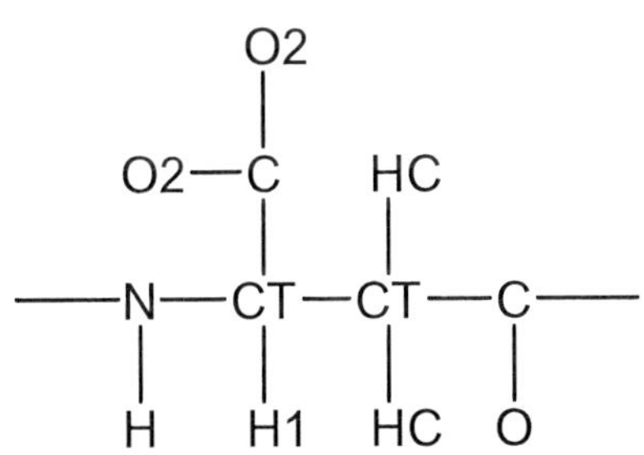

Figure 5. Atom type of the aspartic acid residue in the ff99SB force field.

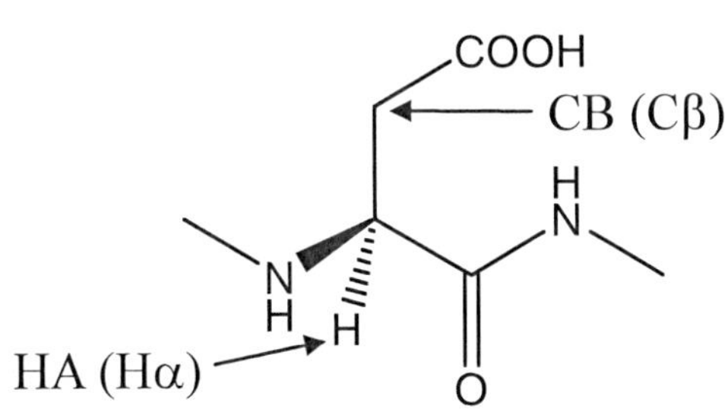

Figure 6. Cβ and Hα of the aspartic acid residue. D-Aspartic acid is constructed from L-Asp by interchanging these two atoms.

2.1.1. Test Peptides

To test the validity of using L-aspartic acid parameters for D-aspartic acid, the peptide corresponding to amino acids 70–88 (KFVIFLDVKHFSPEDLTVK) of human αA-crystallin was used [36]. As mentioned above, the stereoinversion of aspartic acid residues in αA-crystallin has been experimentally observed, and this αA-crystallin 70–88 peptide was the test peptide used in these investigations of the physicochemical properties of αA-crystallin. Two types of αA-crystallin 70–88 peptides were computationally constructed for our tests: one that includes L-Asp76 and the other includes the corresponding D-isomer. MD simulations for both the constructed peptides were performed, and the structural properties obtained from the simulations were compared with experimental retention times. The retention times obtained by reversed-phase high-performance liquid chromatography (RP-HPLC) experiments were 95.2 min for the L-Asp76-containing peptide and 93.2 min for its D-Asp76 counterpart [36]. Because the initial structures of the peptides were constructed by tleap, the initial conformations of the main-chain atoms gave linear structures.

2.1.2. Methods for Test Calculation

To prepare MD simulations, structural minimizations of the added hydrogen atoms were performed first, followed by minimizations of the side chains. Next, whole structures were optimized, and the resulting optimized structures were then used for MD simulations. Minimizations for hydrogen atoms and side chains were performed in vacuo because of the limitations of AMBER 10, and whole molecule minimizations were performed in implicit solvents. For implicit solvent models, analytical linearized Poisson–Boltzmann (ALPB) [37,38] was used. The effective electrostatic size for ALPB was determined to be 20 Å from the result of a utility *elsize* calculation. The effective Born radii for ALPB were obtained from GBn [39]. For implicit solvent models, salt concentrations for ion screening of interactions by the Debye–Hückel limiting law were set to 0.1 M. The Bondi atomic radii were used. The maximum number of cycles was set at 500, 10,000, and 30,000 for minimization of hydrogen atoms, side chains, and whole molecules, respectively. Minimizations were terminated when the energy gradient was less than 0.1 kcal mol^{-1} $Å^{-1}$. For MD simulations, 100-ns simulations were performed in totally using a 1-fs time step at 300 K with the abovementioned implicit solvent model. The covalent bonds containing hydrogen atoms were restricted by SHAKE [40]. The cutoff length of nonbonding interactions was set to 999 Å, and the maximum distance between atom pairs to be considered in pairwise summation, which was used in calculating the effective Born radii, was also equal to 999 Å. Because the interatomic distances never exceeded 999 Å throughout the calculations, these cutoff values were equivalent to no cutoff. For minimizations and MD simulations, the SANDER module of AMBER 10 was used.

After completing MD simulations, structural properties of the two peptides were analyzed. The amino acid sequence, hydrophobicity, and molecular size of the peptides were already known to influence the retention time [41]. Although values for the regression coefficients vary using different HPLC conditions, qualitative tendencies based on specific properties of the peptides are known: larger size and reduced hydrophobicity decrease the retention time [42]. Thus, in this chapter, percent polar surface area (PSA) and molecular volume were calculated, and these values were qualitatively compared with experimental retention times. Because a high correlation between percent PSA and log P was reported [43], percent PSA was used as the indicator for hydrophobicity. PSA was calculated using the DMS program [44,45], and the percentage of PSA for the whole surface was defined as percent PSA. The log P values were obtained from percent PSA using the regression expression derived from the data in ref. [43]. The molecular volume was calculated using the Mol_Volume program [46]. For these calculations, the probe radius was set to 1.4 Å. 3D structures of peptides were extracted from MD trajectories after 50 ns, and the average values were determined for both percent PSA and molecular volume.

2.1.3. Results of Test Calculations

To investigate the practicality of using L-aspartic acid parameters for D-aspartic acid residues, a comparison between the retention times obtained from MD simulations and those acquired experimentally by RP-HPLC [36] was carried out for the αA-crystallin 70–88 peptides. In Table 1, molecular volumes and percent PSA of calculated peptide structures obtained by MD simulations are shown. The values are the ensemble average of MD trajectories after 50 ns. The log P values predicted by the regression expression derived from the data in ref. [43] are also provided. As shown in the table, the molecular volume of the peptide containing D-Asp76 was larger than that of the peptide with L-Asp76, and the hydrophobicity of the D-Asp76 peptide was lower than that of the L-Asp76 peptide. These changes in molecular properties were caused by structural modifications induced by the stereoinversion of the Asp76 residue. Because a larger molecular volume and lower hydrophobicity tend to decrease the retention time by RP-HPLC [41], the molecular properties shown in Table 1 are qualitatively consistent with the experimental results shown in ref. [36]. These results indicate that L-aspartic acid parameters are appropriate for D-aspartic acid residues. In addition, the results also suggest the utility of MD simulations for the prediction of RP-HPLC retention times.

For a quantitative prediction of the retention time, a large number of experimental data obtained under specific HPLC conditions are required because regression expressions must be derived from experimental data observed under distinct conditions.

Table 1. Molecular volume, percent PSA, and predicted log P of αA-crystallin 70–88 peptides

	molecular volume / Å^3	percent PSA	log P[a]
L-Asp76 including peptide	5037	26.5%	−0.671
D-Asp76 including peptide	5165	27.0%	−0.743

[a] Calculated by regression analysis of the data in ref. [43].

As shown in ref. [42], significantly divergent regression coefficients are derived with different HPLC conditions. In addition, in the amino-acid-sequence-dependent term included in the regression expression (Sum_{AA} term of ref. [41]), there is no parameter for D-aspartic acid residues. Therefore, a large number of experimental data for peptides with D-aspartic acid residues obtained under specific HPLC conditions are indispensable for quantitative predictions of retention time; however, only limited data have been obtained for these peptides at present. On the other hand, it appears that qualitative predictions of retention times are possible using MD simulations because MD results were consistent with experimental data. In the future, quantitative regression expressions are expected to be derived from experimental HPLC data and MD results when sufficient empirical data is available. Retention time predictions for these peptides cannot be performed by the method using only the amino acid sequence [47] because peptide conformations affect retention times. In addition, the retention time prediction using Gibbs energy calculations [48] requires huge computational costs. Thus, the regression expression derived from MD simulations and experimental data are expected to be useful for retention time predictions of peptides containing D-aspartic acid residues.

2.2. General Amber Force-Field Parameters

In this section, GAFF parameters are used for D-aspartic acid residues. Because GAFF is a widely used force field for small molecules, it is expected to be suitable for simulations of short peptides such as enzyme substrates. For these simulations, atom types and atomic charges are generally assigned by the Antechamber tool. The tool can automatically prepare AMBER calculations using semiempirical quantum chemical calculations to obtain atomic charges and identify the atom types in molecules.

2.2.1. Test Set

For calculations with a short peptide containing a D-aspartic acid residue, substrates of the protein L-isoaspartyl (D-aspartyl) *O*-methyltransferase (PIMT) [49-51] were used. PIMT is the enzyme that repairs isomerized aspartic acid residues in living body, and it requires *S*-adenosyl-L-methionine (SAM) or *S*-adenosyl-L-homocysteine (SAH) as a cofactor. Human PIMT recognizes D-α-aspartic acid or L-β-aspartic acid residues in proteins or peptides.

Methylation of the carboxyl moiety in the side chain of isomerized aspartic acid residues is catalyzed by PIMT, and these methylated residues readily transform to the cyclic intermediate described earlier. In the enzymatic reaction, SAM acts as the methyl source and is converted to SAH in the process (Figure 7).

The binding affinities of peptides containing D-α-aspartic acid and L-β-aspartic acid residues for human PIMT are reportedly different. The affinity of the peptide containing L-β-aspartic acid residue was ~700–10000 times higher than that of the D-α-aspartic acid residue, although the amino acid sequences were the same [52,53]. Because three-dimensional structures of human PIMT–substrate complexes have not yet been experimentally observed, the reasons for the difference are unknown. Thus, MD simulations of PIMT complexes are expected to be useful.

In test calculations, the PIMT–peptide substrate complex structures were predicted using computational docking and MD simulations. The peptide substrates are the hexapeptides Val–Tyr–Pro–Asp–His–Ala. Two variations were used: a peptide with L-β-Asp4 (substrate *1*) and one with D-α-Asp4 (substrate *2*). Both peptides are known substrates for PIMT [52-55]. The three-dimensional structures of these hexapeptides were constructed by Discovery Studio 2.1 [56]. The structure of PIMT was extracted from the Protein DataBank (PDB) [57] (PDB ID: 1i1n [58]). The missing atoms of 1i1n were added, and SAH in 1i1n was changed to SAM.

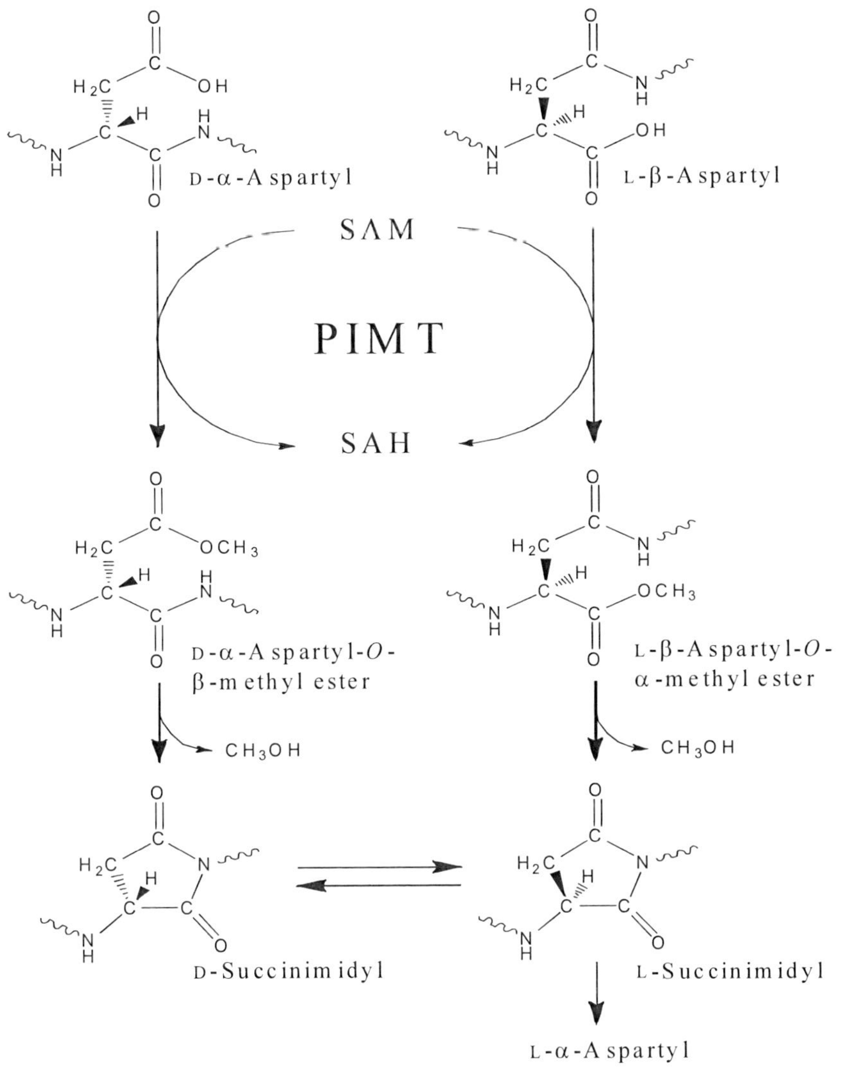

Figure 7. Catalytic reaction of PIMT.

2.2.2. Methods for Test Calculations

Initially, structural minimization of PIMT was performed. A 500-step minimization was performed for SAM alone, and a 1000-step minimization was done for the whole PIMT–SAM complex. Both minimizations were performed in vacuo. Following the minimization of the PIMT–SAM complex, explicit water molecules were added. For the solvent model, the TIP3P water solvent model [59] with a thickness of 8 Å was used. Then, a 1000-step minimization only for water molecules and a 2500-step minimization for the whole system were performed. For the minimized system, the substrate of *P. furiosus* PIMT whose structure was extracted from PDB ID: 1jg3 [60] was added, and a minimization of the whole system was performed. After the minimization, the substrate was removed. From these procedures, a substrate-binding site in PIMT was prepared.

For the minimized PIMT–SAM complex, computational docking of peptide substrates was performed. As mentioned in section 2.1.2, two types of substrates were used for docking: VYP-L-β-Asp-HA (substrate *1*) and VYP-D-Asp-HA (substrate *2*). For calculations, LibDock [61] implemented in Discovery Studio 2.1 was used. The ligand binding site was detected by Discovery Studio 2.1, and site 1, which was located near the cofactor SAM, was selected. The radius of the ligand-binding site for a hotspots search was set to 15 Å. The "number of hotspots" parameter was equal to 1000, and the "tolerance" was set to 0.3 Å and 0.5 Å. The parameter "conformation method" was "BEST." The default settings were used for other parameters. From the docking results, the docking poses of the human PIMT–SAM–substrate complexes were screened by comparing the conformations and orientations of the ligands to the PIMT complex of *P. furiosus*. MD simulations of the selected PIMT–SAM–substrate complexes were then performed.

For MD simulations, the TIP3P water solvent model with a thickness of 8 Å was used. Following the preparatory minimizations performed in the same way as described above, 20-ns MD simulations with a 2.0-fs time step at 300 K were performed. The covalent bonds containing hydrogen atoms were restricted by SHAKE [40]. The force-field parameters for PIMT and TIP3P were ff03, and those for SAM and peptide substrates were GAFF. The system including an explicit solvent model was simulated with a periodic boundary condition, and the particle mesh Ewald method [62] was employed to treat long-range electrostatic interactions. A cutoff of 10 Å was used for nonbonding interactions.

2.2.3. Results of Test Calculations

For the docking calculations, 100 poses were obtained for substrate *1* at each of the two tolerance levels (0.3 and 0.5). The number of docking poses for substrate *2* was also approximately 100. LibDock can predict complex structures in the same manner as protein-small ligand complexes even though peptide substrates were used in test calculations. As before, the docking poses were screened by comparing the conformations and orientations of the ligands to the PIMT complex of *P. furiosus*, and MD simulations were then performed. The representative PIMT–SAM–substrate complex structures are illustrated in Figure 8. As shown in the figure, the side chain of the aspartic acid residue for both substrates *1* and *2* was located near the cofactor SAM in PIMT. The location was consistent with the catalytic mechanism of PIMT shown in Figure 7.

This could originate from the fact that the hydrogen bonds involving the side-chain carboxyl groups of these residues are fundamental for the substrate recoginition by PIMT, and the remaining functionality of the aspartic acid residue does not seem to be important. The diversity of the substrate recognitions by PIMT can be attributed to the fact that crucial hydrogen bonds engage L-β-aspartic acid or D-α-aspartic acid residues themselves and not other residues. On the other hand, there were some differences between substrates *1* and *2* regarding hydrogen-bonding patterns. On the C-terminal side of the aspartic acid residue of substrate *1*, hydrogen bonds were formed between the main chain of L-β-Asp and that of Ala61 and between the main chain of His and that of Ile212, although these hydrogen bonds were not observed for substrate *2*.

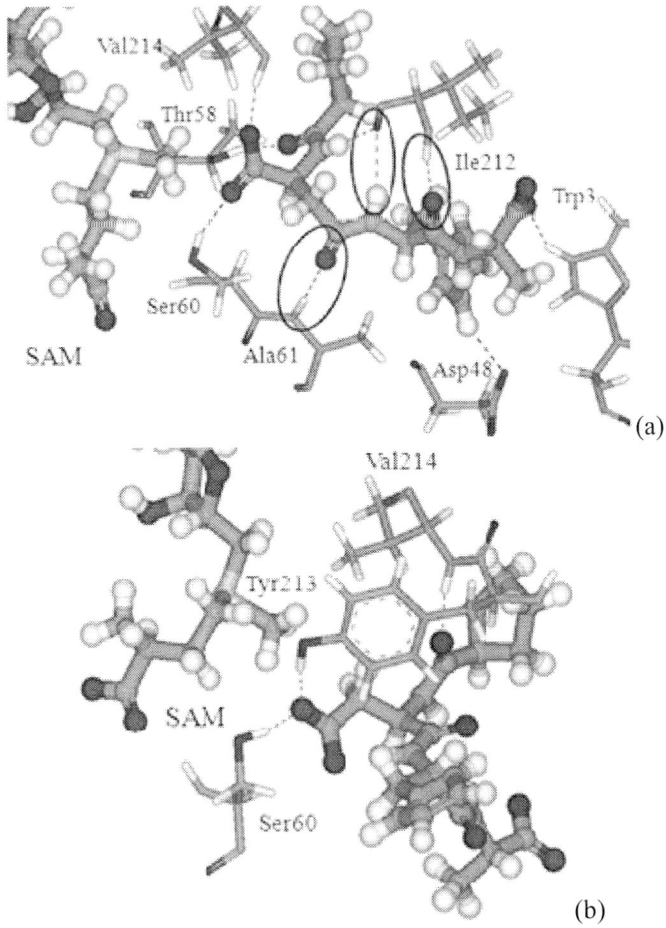

Figure 8. PIMT–SAM–substrate complex structures. Additional hydrogen bonds on the C-terminal side of aspartic acid are circled. (a) PIMT–SAM–substrate *1* (b) PIMT–SAM–substrate *2*.

The binding affinity between human PIMT and substrate *1* is higher than that between human PIMT and substrate *2*, and these C-terminal side hydrogen bonds may contribute to this enhanced binding. Because the importance of the C-terminal side of aspartic acid residues for recognition of PIMT substrates has been reported [53], these computational results were consistent with the experimental results. In addition, more intramolecular hydrogen bonds were observed for substrate *2* than for substrate *1*. This may lead to weaker intermolecular hydrogen bonds. Furthermore, strong intramolecular hydrogen bonds may cause an entropy loss, and these entropic effects might be responsible for the low binding affinity of substrate *2*. Hydrophobic effects play important roles in ligand recognition by proteins. As shown in Figure 8(a), Ile212 and Val214 of PIMT were located near Pro3 of substrate *1*. Because it has been previously reported that hydrophobic groups are favored for the (n − 1) residue, the results showing that (n − 1) residue Pro3 of substrate *1* was surrounded by hydrophobic residues were also consitent with experimental data. On the other hand, the distances between Pro3 of substrate *2* and hydrophobic residues of PIMT were greater than those of substrate *1*.

Thus, hydrophobic effects may contribute to the differences in the binding affinities between substrates *1* and *2*.

Structural changes to proteins induced by substrate docking can be simulated by MD. The changes to the PIMT structure are shown in Table 2. Pro50, Thr58, Ser60, Ile212, Tyr213, and Val214 are included in the substrate binding pocket of PIMT, and the distances between these residues are shown. "Initial" refers to the initial structure before the docking calculations. As shown in the table, the distance between Pro50 and Ile212 decreased on binding substrate *1*. These two residues were located on both sides of substrate *1*, and this induced-fit structural change may assist in the substrate recognition by PIMT. In addition, the distances between Ser60 and Tyr213 and between Thr58 and Val214 also decreased, and the substrate binding site of PIMT acquired the "closed form" on binding substrate *1*.

On the other hand, structural changes to the PIMT–SAM–substrate *2* complex were different from those of PIMT–SAM–substrate *1*. The substrates of PIMT are peptides and proteins because PIMT repairs isomerized aspartic acid residues in both. Thus, protein–protein complexes are formed by substrate recognition by PIMT. Generally, induced fit plays an important role in such complexes; therefore, analyses of structural changes are indispensable to the investigation of mechanisms of substrate recognition by PIMT. For this situation, simple docking calculations are insufficient, and MD simulations are needed.

Table 2. Interatomic distances (Å) between residues around the active site of PIMT

	Initial	PIMT-substrate *1* complex	PIMT-substrate *2* complex
Pro50 Cγ–Ile212 Cγ	9.58	7.21	13.95
Ser60 Cβ–Tyr213 Cβ	8.29	7.68	9.45
Thr58 Cβ–Val214 Cβ	6.50	6.22	7.63

MD simulations of L-β-aspartic acid and D-α-aspartic acid residues using GAFF can reproduce the substrate recognition by PIMT, and the results indicate that the difference in binding affinities between substrates *1* and *2* is caused by differences in the hydrogen-bonding patterns and hydrophobic effects. The results suggest that GAFF is suitable for studying L-β-aspartic acid and D-α-aspartic acid residues in short peptides.

CONCLUSION

In this manuscript, the feasibility of using L-aspartic acid parameters for D-aspartic acid residues in computational studies of comparatively longer peptides and using GAFF for simulations with D-aspartic acid residues in shorter peptides was validated using test calculations. D-Aspartic acid residues in proteins and peptides are considered to play important roles in age-related diseases. Thus, these structural bioinformatics studies of proteins that incorporate D-aspartic acid residues using appropriate force-field parameters are expected to be applicable to basic and applied biosciences such as drug design.

REFERENCES

[1] Fujii, N.; Saito, T. (2004). Homochirality and life. *Chem. Rec., 4*, 267–278.

[2] Takahashi, O.; Kobayashi, K.; Oda, A. (2010). Modeling the enolization of succinimide derivatives, a key step of racemization of aspartic acid residues: importance of a two-H_2O mechanism. *Chem. Biodiversity, 7*, 1349–1356.

[3] Takahashi, O.; Kobayashi, K.; Oda, A. (2010). Computational modeling of the enolization in a direct mechanism of racemization of the aspartic acid residue. *Chem. Biodiversity, 7*, 1630–1633.

[4] Fujii, N.; Tajima, S.; Tanaka, N.; Fujimoto, N.; Takata, T.; Shimo-Oka, T. (2002). The presence of D-β-aspartic acid-containing peptides in elastic fiber of sun-damaged skin: a potent marker for ultraviolet-induced skin aging. *Biochem. Biophys. Res. Commun., 294*, 1047–1051.

[5] Roher, A. E.; Lowenson, J. D.; Clarke, S.; Wolkow, C.; Wang, R.; Cotter, R. J.; Reardon, I. M.; Zürcher-Neely, H. A.; Heinrikson, R. L.; Ball, M. J.; Greenberg, B. D. (1993). Structural alterations in the peptide backbone of beta-amyloid core protein may account for its deposition and stability in Alzheimer's disease. *J. Biol. Chem., 268*, 3072–3083.

[6] Tomiyama, T.; Asano, S.; Furiya, Y.; Shirasawa, T.; Endo, N.; Mori, H. (1994). Racemization of Asp23 residue affects the aggregation properties of Alzheimer amyloid beta protein analogues. *J. Biol. Chem., 269*, 10205–10208.

[7] Sakai-Kato, K.; Naito, M.; Utsunomiya-Tate, N. (2007). Racemization of the amyloidal β Asp^1 residue blocks the acceleration of fibril formation caused by racemization of the Asp^{23} residue. *Biochem. Biophys. Res. Commun., 364*, 464–469.

[8] Fujii, N.; Satoh, K.; Harada, K.; Ishibashi, Y. (1994). Simultaneous stereoinversion and isomerization at specific aspartic acid residues in alpha A-crystallin from human lens. *J. Biochem., 116*, 663–669.

[9] Sim, E.; Lack, N.; Wang, C.; Long, H.; Westwood, I.; Fullam, E.; Kawamura, A. (2008). Arylamine *N*-acetyltransferases: structural and functional implications of polymorphisms. *Toxicology, 254*, 170–183.

[10] [Wang, L.; Weinshilboum, R. (2006). Thiopurine *S*-methyltransferase pharmacogenetics: insights. challenges and future directions *Oncogene, 25*, 1629–1638.

[11] Human Cytochrome P450 (CYP) Allele Nomenclature Committee. http://www.cypalleles.ki.se/

[12] Weiner, S. J.; Kollman, P. A.; Case, D. A.; Singh, U. C.; Ghio, C.; Alagona, G.; Profeta, S.; Weiner, P. (1984). A new force field for molecular mechanical simulation of nucleic acids and proteins. *J. Am. Chem. Soc., 106*, 765–784.

[13] Weiner, S. J.; Kollman, P. A.; Nguyen, D. T.; Case, D. A. (1986). An all atom force field for simulations of proteins and nucleic acids. *J. Comput. Chem., 7*, 230–252.

[14] Cornell, W. D.; Cieplak, P.; Bayly, C. I.; Gould, I. R.; Merz, K. M.; Ferguson, D. M.; Spellmeyer, D. C.; Fox, T.; Caldwell, J. W.; Kollman, P. A. (1995). A second generation force field for the simulation of proteins, nucleic acids, and organic molecules. *J. Am. Chem. Soc., 117*, 5179–5197.

[15] Wang, J.; Cieplak, P.; Kollman, P. A. (2000). How well does a restrained electrostatic potential (RESP) model perform in calculating conformational energies of organic and biological molecules? *J. Comput. Chem.*, *21*, 1049–1074.
[16] Hornak, V.; Abel, R.; Okur, A.; Strockbine, B.; Roitberg, A.; Simmerling, C. (2006). Comparison of multiple Amber force fields and development of improved. *Proteins: Struct. Funct. Bioinf.*, *65*, 712–725.
[17] [Duan, Y.; Wu, C.; Chowdhury, S.; Lee, M. C.; Xiong, G.; Zhang, W.; Yang, R.; Cieplak, P.; Luo, R.; Lee, T. (2003). A point-charge force field for molecular mechanics simulations of proteins based on condensed-phase quantum mechanical calculations. *J. Comput. Chem.*, *24*, 1999–2012.
[18] Lee, M. C.; Duan, Y. (2004). Distinguish protein decoys by using a scoring function based on a new Amber force field, short molecular dynamics simulations, and the generalized Born solvent model. *Proteins: Struct. Funct. Genet.*, *55*, 620–634.
[19] Simmerling, C.; Strockbine, B.; Roitberg, A. E. (2002). All-atom structure prediction and folding simulations of a stable protein.*J. Am. Chem. Soc.*, *124*, 11258–11259.
[20] Oda, A.; Kobayashi, K.; Takahashi, O. (2010). Computational study of the three-dimensional structure of, *N*-acetyltransferase 2–acetyl coenzyme A complex. *Biol. Pharm. Bull.*, *33*, 1639–1643.
[21] Senapati, S.; Bui, J. M.; McCammon, J. A. (2005). Induced fit in mouse acetylcholinesterase upon binding a femtomolar inhibitor: A molecular dynamics study. *J. Med. Chem.*, *48*, 8155–8162.
[22] Oda, A.; Yamaotsu, N.; Hirono, S. (2004). Studies of binding modes of (S)-mephenytoin to wild types and mutants of cytochrome P450 2C19 and 2C9 using homology modeling and computational docking. *Pharm. Res.*, *21*, 2270–2278.
[23] Oda, A.; Yamaotsu, N.; Hirono, S. (2005). New AMBER force-field parameters of heme iron for cytochrome P450s determined by quantum chemical calculations of simplified models. *J. Comput. Chem.*, *26*, 818–826.
[24] Allinger, N. L. (1977). Conformational analysis. 130. MM2. A hydrocarbon force field utilizing V1 and V2 torsional terms. *J. Am. Chem. Soc.*, *99*, 8127–8134.
[25] Halgren, T. A. (1996). Merck molecular force field. I. Basis, form, scope, parameterization, and performance of MMFF94. *J. Comput. Chem.*, *17*, 490–519.
[26] Jorgensen, W. L.; Tirado-Rives, J. (1988). The OPLS force field for proteins. energy minimizations for crystals of cyclic peptides and crambin. *J. Am. Chem. Soc.*, *110*, 1657–1666.
[27] Jorgensen, W. L.; Maxwell, D. S.; Tirado-Rives, J. (1996). Development and testing of the OPLS all-atom force field on conformational energetics and properties of organic liquids. *J. Am. Chem. Soc.*, *118*, 11225–11236.
[28] Brooks, B. R.; Bruccoleri, R. E.; Olafson, B. D.; States, D. J.; Swaminathan, S.; Karplus, M. (1983). CHARMM: A program for macromolecular energy, minimization, and dynamics calculations. *J. Comput. Chem.*, *4*, 187–217.
[29] Scott, W. R. P.; Hünenberger, P. H.; Tironi, I. G.; Mark, A. E.; Billeter, S. R.; Fennen, J.; Torda, A. E.; Huber, T.; Krüger, P.; van Gunsteren, W. F. (1999). The GROMOS biomolecular simulation package. *J. Phys. Chem. A*, *103*, 3596–3607.
[30] Case, D. A.; Darden, T. A.; Cheatham, T. E.; Simmerling, C. L.; Wang, J.; Duke, R. E.; Luo, R.; Crowley, M.; Walker, R. C.; Zhang, W.; Merz, K. M.; Wang, B.; Hayik, S.; Roitberg, A.; Seabra, G.; Kolossváry, I.; Wong, K. F.; Paesani, F.; Vanicek, J.; Wu, X.;

Brozell, S. R.; Steinbrecher, T.; Gohlke, H.; Yang, L.; Tan, C.; Mongan, J.; Hornak, V.; Cui, G.; Mathews, D. H.; Seetin, M. G.; Sagui, C.; Babin, V.; Kollman, P. A. (2008) *AMBER 10*, San Francisco, CA: University of California.

[31] Kobayashi, K.; Oda, A.; Takahashi, O. (2010). Evaluation of molecular dynamics simulation with explicit solvent using bovine pancreatic trypsin inhibitor. *J. Tohoku Pharm. Univ.*, *57*, 67–72.

[32] Wang, J.; Wolf, R. M.; Caldwell, J. W.; Kollamn, P. A.; Case, D. A. (2004). Development and testing of a general Amber force field. *J. Comput. Chem.*, *25*, 1157–1174.

[33] Wang, J.; Wang, W.; Kollman, P. A.; Case, D. A. (2006). Automatic atom type and bond type perception in molecular mechanical calculations. *J. Mol. Graph. Modell.*, *25*, 247–260.

[34] Oda, A.; Kobayashi, K.; Takahashi, O. (2011). Comparison of molecular dynamics simulation methods for amyloid β_{1-42} monomers containing D-aspartic acid residues for predicting retention times in chromatography. *J. Chromat. B*, submitted for publication.

[35] Noji, I.; Oda, A.; Kobayashi, K.; Takahashi, O. (2011). Computational investigation of the substrate recognition mechanism of protein D-aspartyl (L-isoaspartyl) *O*-methyltransferase by docking and molecular dynamics simulation studies and application to interpret size exclusion chromatography data. *J. Chromat. B*, in press.

[36] Fujii, N.; Fujii, N.; Kida, M.; Kinouchi, T. (2010). Influence of Lβ-, Dα- and Dβ-Asp isomers of the Asp-76 residue on the properties of αA-crystallin 70–88 peptide. *Amino Acids*, *39*, 1393–1399.

[37] Sigalov, G.; Scheffel, P.; Onufriev, A. (2005). Incorporating variable environments into the generalized Born model. *J. Chem. Phys.*, *122*, 094511.

[38] Sigalov, G.; Fenley, A.; Onufriev, A. (2006). Analytical electrostatics for biomolecules: Beyond the generalized Born approximation. *J. Chem. Phys.*, *124*, 124902.

[39] Mongan, J.; Simmerling, C.; McCammon, J. A.; Case, D. A.; Onufriev, A. (2006). Generalized Born with a simple, robust molecular volume correction. *J. Chem. Theory Comput.*, *3*, 156–169.

[40] Ryckaert, J.-P.; Ciccotti, G.; Berendsen, H. J. C. (1977). Numerical integration of the cartesian equations of motion of a system with constraints: Molecular dynamics of *n*-alkanes. *J. Comput. Phys.*, *23*, 327–341.

[41] Kaliszan, R.; Bączek, T.; Cimochowska, A.; Juszczyk, P.; Wiśniewska, K.; Grzonka, Z. (2005). Prediction of high-performance liquid chromatography retention of peptides with the use of quantitative structure-retention relationships. *Proteomics*, *5*, 409–415.

[42] Bączek, T.; Wiczling, P.; Marszałł, M.; Heyden, Y. V.; Kaliszan, R. (2005). Prediction of peptide retention at different HPLC conditions from multiple linear regression models. *J. Proteome Res.*, *4*, 555–563.

[43] Barlow, D.; Satoh, T. (1994). The design of peptide analogues for improved absorption. *J. Contr. Release*, *29*, 283–291.

[44] Richards, F. M. (1977). Areas, volumes, packing, and protein structure. *Annu. Rev. Biophys. Bioeng.*, *6*, 151–176.

[45] Huang, C. (2002) *DMS*, San Francisco, CA: University of California.

[46] Balaeff, A. (2001). *Mol_Volume*, Urbana-Champaign, IL: University of Illinois.

[47] Meek, J.L. (1980). Prediction of peptide retention times in high-pressure liquid chromatography on the basis of amino acid composition. *Proc. Natl. Acad. Sci. USA*, *77*, 1632–1636.

[48] Tsai, C.; Liu, C.; Chan, Y.; Tsai, H. G.; Ruaan, R. (2010). Study of conformation effects on the retention of small peptides in reversed-phase chromatography by thermodynamic analysis and molecular dynamics simulation. *J. Phys. Chem. B*, *114*, 11620–11627.

[49] Furuchi, T.; Homma, H. (2007). The role of isomerized protein repair enzyme, PIMT, in cellular functions. *Yakugaku Zasshi*, *127*, 1927–1936.

[50] O'Connor, C. M. (2006). In F. Tamanoi, S. Clarke (Eds.), *The Enzymes*, *Vol. 24* (p. 385). San Diego, CA: Academic of Dermatology.

[51] Rutherford, K.; Daggett, V. (2009). The V119I polymorphism in protein *L*-isoaspartate *O*-methyltransferase alters the substrate-binding interface. *Protein Eng. Des. Sel.*, *22*, 713–721.

[52] Lowenson, J. D.; Clarke, S. (1992). Recognition of D-aspartyl residues in polypeptides by the erythrocyte L-isoaspartyl/D-aspartyl protein methyltransferase. Implications for the repair hypothesis. *J. Biol. Chem.*, *267*, 5985–5995.

[53] Thapar, N.; Griffith, S. C.; Yeates, T. O.; Clarke, S. (2002). Protein repair methyltransferase from the hyperthermophilic archaeon *Pyrococcus furiosus* : Unusual methyl-accepting affinity for D-aspartyl and *n*-succinyl-containing peptides. *J. Biol. Chem.*, *277*, 1058–1065.

[54] Thapar, N.; Clarke, S. (2000). Expression, purification, and characterization of the protein repair L-isoaspartyl methyltransferase from *Arabidopsis thaliana*. *Protein Express. Purif.*, *20*, 237–251.

[55] DeVry, C. G.; Clarke, S. (1999). Polymorphic forms of the protein L-isoaspartate (D-aspartate) *O*-methyltransferase involved in the repair of age-damaged proteins. *J. Hum. Genet.*, *44*, 275–288.

[56] Accelrys (2008). *Discovery Studio 2.1*, San Diego, CA: Accelrys.

[57] Berman, H. M.; Westbrook, J.; Feng, Z.; Gilliland, G.; Bhat, T. N.; Weissig, H.; Shindyalov, I. N.; Bourne, P. E. (2000). The protein data bank. *Nucleic Acids Res.*, *28*, 235–242.

[58] Smith, C. D.; Carson, M.; Friedman, A. M.; Skinner, M. M.; Delucas, L.; Chantalat, L.; Weise, L.; Shirasawa, T.; Chattopadhyay, D. (2002). Crystal structure of human L-isoaspartyl-*O*-methyl-transferase with *S*-adenosyl homocysteine at 1.6-Å resolution and modeling of an isoaspartyl-containing peptide at the active site. *Protein Sci.*, *11*, 625–635.

[59] Jorgensen, W. L.; Chandrasekhar, J.; Madura, J.; Klein, M. L. (1983). Comparison of simple potential functions for simulating liquid water. *J. Chem. Phys.*, *79*, 926–935.

[60] Griffith, S.C.; Sawaya, M. R.; Boutz, D. R.; Thapar, N.; Katz, J. E.; Clarke, S.; Yeates, T. O. (2001). Crystal structure of a protein repair methyltransferase from *Pyrococcus furiosus* with its L-isoaspartyl peptide substrate. *J. Mol. Biol.*, *313*, 1103–1116.

[61] Diller, D. J.; Merz, K. M. (2001). High throughput docking for library design and library prioritization. *Proteins: Struct. Funct. Genet.*, *43*, 113–124.

[62] Darden, T.; York, D.; Pedersen, L. (1993). Particle mesh Ewald–an Nlog(N) method for Ewald sums in large systems. *J. Chem. Phys.*, *98*, 10089–10092.

In: Tyrosine and Aspartic Acid
Editors: J. E. Jones and D. M. Morano
ISBN: 978-1-62100-752-4

Chapter 6

CENTRAL FUNCTIONS OF D-ASPARTIC ACID

R. C. Cervantes Cianca**, L. R. Faro, R. Durán and Miguel Alfonso

Department of Functional Biology and Health Sciences, Faculty of Biology, University of Vigo, Campus As, Lagoas-Marcosende, 36310 Vigo, Spain

ABSTRACT

Over the past years, it has been demonstrated that amino acids and some neurotransmitters –serotonin and dopamine– are responsible for the most of synaptic communication in our brain. Along the time, research has been basically focused to L-enantiomers of amino acids. L-Amino acids build blocks of proteins; each protein is composed of a chain made up of amino acids in a unique sequence. Prior to the 80´s decade, L-amino acids were considered as vital for different functions in living organisms. However, due to the recent advances of analytical techniques and methods, the presence of D-amino acids has been so far recognized in the biological world with essential physiological functions. Besides these advances in analytical techniques, it has also allowed a more accurate measurement of small amounts of D-amino acids among much larger pools of L-amino acids in biological fluids and tissues (Friedman and Levin, 2011; Ilisz et al. 2010; Kirschner and Green, 2009). For example, D-Aspartic acid (D-Asp) –considered as one of the most studied D-amino acids– is an endogenous amino acid found in nervous and endocrine tissues of invertebrates and vertebrates. D-Asp promotes the proteins synthesis in the development of the nervous system and regulates the hormonal release in the endocrine glands. D-Asp –which acts as a neurotransmitter or neuromodulator– is one of the major excitatory amino acids in the brain. Consequently, for a better understanding of its biological importance, it is crucial to study their occurrence and metabolism in living systems. In this work, the study will be focused in the role of D-Asp in central nervous system.

* Email: rosacccianca@gmail.com.

1. Introduction

All the amino acids found in proteins, are of α-configuration (except glycine which is not optically active). The L-prefix attached to these amino acids, is used to refer to the configuration of the α-carbon atom. But, in the nature other non proteinogenic amino acids exist, which have different dispersion of the light. They are the D-amino acids (D-form) (D'Aniello, 2007a; Greenstein et al. 1953).

D-amino acids generally were viewed as natural occurrence (Felbeck and Wiley, 1987). Until the 1970s D-amino acids were not considered essential for any vital function in higher organisms (Spinelli et al. 2006). D-amino acids occur, both freely dissolved, and incorporated into peptides and metabolites in the tissues of animals and plants (D'Aniello et al. 1993b). It has been suggested that the D-isomers of amino acids were as common as the L-forms in ancient prebiotic times on earth (Felbeck and Wiley, 1987). Aono and Yuasa (1977) supposed that the supply of L-amino acids was maintained by chemical racemization from the D-isomers left behind. Katane and Homma (2011) point out that in 1986, Dunlop et al. reported for the first time the existence of free D-Asp in mammals, including humans. In the 70's decade, it became clear that in a variety of lower animals, such as bacteria and insects, some D-amino acids were indeed present (Katane and Homma, 2011). Over the two last decades, evidence has been accumulated indicating that some D-amino acids occur and fulfill specific biological functions in mammals including humans (Fisher, 1998; Fuchs et al. 2005; Hamase et al. 2002). By the way, recent advances in analytical methods (Hamase et al. 1997), for separating chiral amino acids have revealed that living organisms contain several D-amino acids, either free or as components of proteins (Dunlop et al. 1986; Katane and Homma, 2011).

As D-amino acids play biological roles, some alterations in the concentrations of them, might produce some disorders. D-amino acids concentrations may not only help in the diagnostic process, but also provide novel therapeutic targets. In addition, the presence and important roles of D-amino acids in higher organisms do not only challenge former theories on mammalian physiology, but also contribute to exciting new insights in human disease (Fuchs et al. 2005).

In particular, free D-serine (D-Ser) and D-Asp has been identified in a wide variety of mammalian tissues and cells at relatively high concentrations. These two amino acids have been extensively studied about their biological properties.

D-Ser has been studied most extensively and it was shown to play a role in excitatory amino acids metabolism, being a co-agonist of the N-methyl-D-aspartate receptor (NMDAR). Insight into D-Ser metabolism is relevant for physiological NMDAR activation and for all the disorders associated with an altered function of the NMDAR, such as schizophrenia, ischemia, epilepsy, and neurodegenerative disorders (Fuchs et al. 2005). There are different researches about the findings and functions on D-Ser (Hashimoto et al. 1993a; Kumashiro et al. 1995; Nishikawa, 2005; Scolari and Acosta, 2007).

D-Asp is the mostly widespread D-amino acid in animal tissues. Since its discovery in a cephalopod (D'Aniello and Giuditta, 1977), it has been identified in different organisms, from lower animals to mammals and humans (Dunlop et al. 1986; Hashimoto et al. 1993c,d; Homma, 2007) including microorganisms (Nagata et al. 1999c) and plants (Bruckner and Westhauser, 2003). Zagon et al. (1994) mentioned that as well as D-Alanine (D-Ala), D-Asp is part of peptidoglycan structure of bacterial cell walls and is widespread in marine invertebrates. Furthermore, different biochemical and immunohistochemical studies indicate that D-Asp plays a role in both, the nervous and endocrine systems (D'Aniello, 2007b; Homma, 2007). Nevertheless, many details remain unclear. In this communication we present some studies about D-Asp central functions.

2. Central Functions

D-Asp is an endogenous amino acid which has been found in the neuroendocrine tissues of both invertebrates and vertebrates (Topo et al. 2009). As demonstrated by D'Aniello and Giuditta (1977, 1978) and D'Aniello et al. (1995), D-Asp was first detected in the nervous tissues of cephalopods (D'Aniello and Giuditta, 1977) and later in the nervous and endocrine tissues of many other animals including humans (D'Aniello et al. 1993c; D'Aniello and Garcia-Fernandez 2007; Dunlop et al. 1986; Neidle and Dunlop, 1990; Spinelli et al. 2006; Topo et al. 2009).

Different researchers have demonstrated that in chicken embryonic brain (Neidle and Dunlop, 1990), rat brain (Dunlop et al., 1986; Hashimoto et al. 1993d), and human brain (Hashimoto et al. 1993c) very high levels of D-Asp occurs transiently during the last stage of embryonic life or in early postnatal life, suggesting that D-Asp is implicated in the development of the nervous system (D'Aniello G. et al. 2005).

D'Aniello et al. (1996, 1998b), Dunlop et al. (1988), Hashimoto et al. (1993b), Wolosker et al. (2000) as well as Kim et al. (2010), postulate that D-Asp is found in some neuronal populations in the brain and the neuroendocrine tissues; for instance in the catecholaminergic cells of the adrenal medulla, the lobes of the pituitary gland, the pineal gland, and the testes. On other hand, Kim et al. (2010) and Wolosker et al. (2000), statement that in early neonatal stages, high D-Asp densities in the cortical plate, subventricular zone, and discrete portions of the hippocampal formation imply a role in its development. Furthermore Katane and Homma et al. (2011), mentioned that D-Asp is present in the rat brain at about 140 nM/g wet weigh just after birth, but D-Asp decreases to trace levels after three weeks of age (Hashimoto et al. 1995).

By the way, in human fetal cortex, D-Asp is approximately 0.36 μM/g wet weight at 14 weeks gestation (Hashimoto et al. 1993c; Katane and Homma, 2011); the concentration of D-Asp exceeded that of L-Asp; D-Asp levels decreases rapidly to trace levels after birth (Hashimoto et al. 1993c). Errico et al. (2009) point out that this phenomenon is the result of the postnatal onset of D-Asp oxidase expression, the only known enzyme that strictly controls the endogenous levels of D-Asp. Furthermore, within the nervous system this amino acid is concentrated in the axon terminals (synaptosomes) and in synaptic vesicles (Spinelli et al. 2006) together with L-Asp and L-Glu (known like important excitatory neurotransmitters), besides, it is involved in visual activity (D'Aniello, 2005). D'Aniello et al. (2000) have demonstrated that D-Asp plays an important role in hormones release from the neuroendocrine system, where it is capable to induce the release of hypothalamic gonadotropin releasing hormone (GnRH), luteinizing hormone, growth hormone and testosterone. Considering these information, D-Asp may be involved in important cellular regulatory processes during development (Furuchi and Homma, 2005).

Hashimoto et al. (1993c), found high levels of free D-Asp and D-Serin the human fetal frontal cortex at gestational week 14, and trace level at gestational week 41 and postnatal stages. On the other hand, in adolescent and aged individual, they just found half that in the fetuses. They suggested that these D-amino acids might play a pivotal role in cerebral development and functions that are related to the NMDAR. Wolosker et al. (2000), point out that in the brain, D-Asp levels are highest in embryonic and early postnatal stage, furthermore they mentioned that high levels occur in the early postnatal cortical plate and subventricular zone of the cerebral cortical cultures, so that it implies a role in development. They suggest that D-Asp is synthesized from L-Asp in primary cultured neuronal cells of the rat embryo.

Dunlop et al. (1986), Neidle and Dunlop (1990) and Topo et al. (2010b) have reported that in the early stages of the embryonic life of rat and chicken, it has also been detected D-Asp at high concentrations in the brain, indicating a role of D-Asp in the development of the nervous system; as well as in human, D-Asp levels return to lower levels, and remains at that concentration for the rest of the life span of the animal. Whereas, in the adult animal, several findings indicate that D-Asp has also a role in neurotransmission, it may acts like a neurotransmitter or as a neuromodulator (D'Aniello, 2007; Katane and Homma, 2011). About the role of D-Asp in the nervous system, different studies have reported its role in neurotransmission.

D-Asp has been detected in synaptosomes and in synaptic vesicles, where it is released after chemical or electric stimuli. Furthermore, D-Asp increases cAMP in neuronal cells and it is transported from the synaptic clefts to presynaptic nerve cells through a specific transporter (D'Aniello, 2007). Kanai and Hediger (1992) described a high-affinity transport system that is dependent on Na^+ but not Cl^- ions and it is essential for terminating the postsynaptic action by rapidly removing released D-Asp from the synaptic cleft.

Some immunohistochemical approach, using an anti-D-Asp antibody, has revealed D-Asp presence in the rat frontal cortex (D'Aniello, 2007; Schell et al. 1997b; Wang et al. 2000). Schell et al. (1997b), observed in the brain discrete neuronal localizations of D-Asp, especially in the external plexiform layer of the olfactory bulb, hypothalamic supraoptic and paraventricular nuclei and certain brainstem nuclei. They also point out that D-Asp was observed in rats 3–4 weeks old, in septal nuclei and in a subset of stellate and basket cells of the cerebellum. D-Asp also has a direct correlation to light exposure in the retina of some animals (D'Aniello, 2007; Ishida and Gordon, 1981). D'Aniello et al. (2011) point out that D-Asp is able to potentiate the effects of L-Glu in the goldfish retina and has a visual function in the retina of *Sepia officinalis* (D'Aniello S. et al. 2005; Ishida and Gordon, 1981).

D-Asp was found at very high concentrations in the marine mollusk *Aplysia limacine* (Spineli et al. 2006), it was detected in the cerebral ganglia and immunohistochemical studies conducted on cultured neurons using an anti-D-aspartate antibody demonstrated that D-Asp occurs in the soma and dendrites. Anaya and D'Aniello (2006) have reported D-Asp presence as an endogenous compound in the *Branchiostoma lanceolatum* nervous tissue. They reported D-Asp in neural tube at a very high concentration (considering the D/L ratio percentage), as well as in mollusks, while in mammal was detected at relatively low concentration. Errico et al. (2008a) demonstrated that D-Asp has a direct role in regulating mousse hippocampal synaptic plasticity, indicating that increased D-Asp content does not affect basal properties of synaptic transmission but enhances long-term potentiation (LTP) in hippocampal slices from both genetic and pharmacological animal models. On the other hand, Wolosker et al. (2000)

researched about D-Asp ontogeny and its depolarization-induced release with KCl. Holopainen and Kontro (1990) also reported that a high K enhanced D-Asp release from cerebellar astrocytes in primary cultures in the rat.

Other researches, like Furuchi and Homma (2005), mentioned that D-Asp is an agonist for the L-Glu receptor, which may be involved in the development and neurogenesis of the brain, and furthermore concluded that, since the brain is a steroidogenic tissue like the testis and adrenal gland, D-Asp may modulate the production of neurosteroids in the brain, similar to that in latter tissues. Kim et al. (2010) also named that D-Asp may function as a modulator of adult neurogenesis. Most recently, D'Aniello et al. (2011) found that D-Asp prevents long-term depression, attenuate schizophrenia-like symptoms (Errico et al. 2008a), and it is able to rescue hippocampal age-related synaptic plasticity (Errico et al. 2010).

Furthermore, Errico et al. (2008a) adduce that D-Asp enhances LTP in mouse hippocampus. Errico et al. (2010), demonstrated that D-Asp acts as an endogenous NMDAR agonist, since it triggers currents via interaction with each of NR2A-D receptor subunits, furthermore in according to its pharmacological features, they showed that oral administration of D-Asp strongly enhances NMDAR-dependent LTP in adulthood and, in turn, completely rescues the synaptic plasticity decay observed in the hippocampus of aged animals.

Fisher et al. (1991), reported evidence for the presence of D-Asp in the white and gray matter of normal human brains and brains of individuals with Alzheimer's disease (AD); they disclosed that D-Asp occurs at about the same concentration in the gray matter of both normal and Alzheimer brains, whereas in white matter its concentration is more than two times higher in normal than Alzheimer brains. D'Aniello et al. (1998a) point out that NMDAR have been shown to be involved in learning and memory processes. They reported that in Alzheimer's disease, there is a reduction of NMDAR and since D-Asp is an endogenous agonist for the NMDAR, a decrease of endogenous free D-Asp in AD could contribute to the memory deficits observed in this disease.

On other hand, D-Asp has been found to be a selective substrate for the high affinity L-glutamate/L-aspartate transporters throughout the glutamatergic neuronal system (Kanai and Hediger, 1992; Pines et al. 1992).

Whereas, Topo et al. (2010b) point out that in the brain of Alzheimer's patients, D-Asp occurs at a significantly lower concentration than in the brain of healthy subjects belonging to the same age control group. They confirm that D-Asp has a role in cognitive process by its ability to modulate AMPA-like glutamate receptors (Brown et al. 2007). Furthermore D'Aniello et al. (1998) and Topo et al. (2010b) refer that D-Asp is associated with higher cognitive functions, particularly in the frontal cortex, in the parietal, in the occipital, in the cerebellum, as well as in the amygdale and hippocampus. Moreover, Topo et al. (2010b) reported evidence that D-Asp in the rats is involved in learning and memory processes. They demonstrated that rats treated with sodium D-Asp displayed a significant improvement in their spatial learning and memory ability compared with control rats. They mentioned that, though the exact mechanism by which D-Asp induces a significant improvement in learning and memory processes, a plausible explanation could be that in the nervous system D-Asp is localized mainly in synaptic vesicles and it could be a novel neurotransmitter implicated in memory process (Topo et al., 2010b). On the other hand, Scanlan et al. (2010), observed in neuron-containing ganglia compared to surrounding sheath tissues, the preferential synthesis and accumulation of D-Asp over L-Asp; they detected little conversion of D-Asp to L-Asp.

They concluded that their observations point to a role for D-Asp in cell-to-cell signaling with many characteristics similar to classical transmitters.

D'Aniello et al. (2011) considered that the implication of D-Asp in neurological activity could be due to its function as a neurotransmitter. They provided evidence about D-Asp is present in high concentrations in the synaptic vesicles of axon terminals and about its synthesis occurs in neurons. Moreover they probed that D-Asp is released in a Ca^{2+} dependent manner and that their specific receptors for D-Asp are present at the postsynaptic membrane, so that, they concluded that D-Asp fulfills all criteria necessary to be considered a novel endogenous neurotransmitter.

Kim et al. (2010) identified and cloned the mammalian aspartate racemase. This enzyme converts L-Asp to D-Asp, and it is localized with D-Asp in the brain and neuroendocrine tissues. Deletion of the aspartate racemase leads to pronounced defects in adult hippocampal neurogenesis, implicating D-Asp acts as a major regulator of neuronal development. On other hand, Kim et al. (2010) point out that additionally to D-Asp prominence in the developing nervous system; they assume that is implicated in adult hippocampal neurogenesis, so that, they consider that D-Asp may function as a modulator.

Additionally, Errico et al. (2009) have developed an animal model that has highlighted that D-Asp has a neuromodulatory role at N-methyl-D-aspartate receptors (NMDARs) in brain areas where they regulate crucial nervous functions.

In the light of the biological importance of D-Asp at central nervous system and its implications like neurotransmitter and neuromodulator, it is important consider its synthesis and degradation mainly by its implications in health.

Katane and Homma (2011) demonstrated that D-Asp was synthesized in mammalian cells. They used a cell line (PC12 cells) and cultured mammalian cells. D-Asp levels were increased in both, the PC cells and the culture media. Considering that D-Asp was not added to the culture medium and that D-Asp exogenous (Long et al., 1998). Moreover, Long et al. (1998) report that in PC12 cells, the highest D is approximately 14 % and no D-amino acids other than D-Asp are detected. Katane and Homma (2011) also comment that the biosynthesis of D-Asp in mammalian cells is mainly mediated by a specific enzyme racemase, which converts L-Asp to D-Asp in a pyridoxal phosphate (PLP) dependent manner.

On the other hand, D'Aniello et al. (2011) investigated the biochemical origin of D-Asp and the possibility that could be synthesized in nerve tissue. They comment that the conversion of L-Asp to D-Asp by aspartate racemase is the main source generating neuronal D-Asp (D'Aniello et al. 2005; Wolosker et al. 2000). D'Aniello et al. (2011) prepared some synaptosomes, synaptic vesicles (SVs), and postsynaptic membranes (PSMs) from a mammal (*Rattus norvegicus*) and a mollusk (*Loligo vulgaris)* in order to probe their theory. They probed that D-Asp is present in high concentrations in the SVs of axon terminals and that D-Asp is synthesized in neurons by conversion of L-Asp to D-Asp across the enzyme D-aspartate racemase. They concluded that D-Asp is synthesized in the soma of the neuron, and transported along the axon to the nerve terminals, they remark the fact that neither the synaptosomes nor the SVs possess D-aspartate racemase.

On the other hand, Miao et al. (2006) stated that miliMolar concentrations of D-Asp were found in various regions of the CNS of *Aplysia californica.* They also reported a heterogeneous distribution of D-Asp in the neurons tested and confirmed the synthesis of D-Asp from L-Asp.

Friedman and Lavin (2011), remark that D-Asp is the precursor for the important neurotransmitter agonist NMDA. Aspartate racemase is one of only two racemases found in mammals, indicating a native ability to regulate levels D-Asp (Topo et al. 2010a). Furuchi and Homma (2005) and Friedman and Lavin (2011) confirm that the ability of mammalian cells to regulate D-Asp implies that D-Asp is a novel type of messenger in the mammalian body.

Furthermore, Friedman and Lavin (2011) communicate that some studies were led to study the regulation of D-Asp; mice strains were developed to be deficient in D-Asp oxidase (Errico et al. 2006; Huang et al. 2006). D'Aniello et al. (1993) and Zaar et al. (2002) mentioned that D-Asp oxidase is the only enzyme known to metabolize D-Asp. D'Aniello et al. (1993) though that D-Asp oxidase no only oxidizes in vivo exogenous D-Asp, but D-Asp oxidase also has the role to act as detoxifying agents to oxidize endogenous D-Asp that may have accumulated during aging. Zaar et al. (2002) mentioned that D-Asp oxidase can regulate D-Asp levels in different regions of the brain. In addition, Friedman and Lavin (2011) communicate that with the deficiency of D-Asp oxidase, D-Asp was accumulated in the kidney, brain, and spleen at levels 10–26 times normal (Errico et al. 2006).

By the way, with the increased of D-Asp, NMDA also significantly increased in the brain (Friedman and Lavin, 2011). Huang et al. (2006) found that D-Asp is accumulated in the pituitary with concurrent decrease synthesis and levels of pituitary proopiomelanocortin. Friedman and Lavin (2011) consider that an excessive level of D-Asp can have negative effects in the body. Burrone et al. (2010) and Friedman and Lavin (2011) point out that D-Asp activated antioxidative, cytoprotective effects in the kidney and enhanced caspase levels, indicative of apoptosis in brain and heart tissues. Besides D-Asp aggravated nephritis in rats induced by *Staphylococcus aureus* bacteria, and by other hand K and Mg depletion induced by diuretics in rats is inhibited by D-Asp (Friedman and Lavin, 2011; Iezhitsa, 2004; Koyuncuoglu et al. 1988).

Friedman and Lavin (2011) and Fujii and Kaji (2008) comment that the *in vivo* formation of D-Asp is a process of an inversion of configuration of L to D-Asp, and it is possible that the ratio D/L can and do exceed. On other hand, some eye diseases and diabetic may be affected by age related accumulation of D-Asp and glycations end product (Friedman and Lavin, 2011; Kaji, 2010). Even more like D-Asp accumulation has been implicated in the pathogenesis of Alzheimer's disease, as well as arteriosclerosis and cataracts. Friedman and Lavin (2011) raise the question of whether the enzyme D-Asp oxidase (Kinouchi, 2004) can slow down or reverse the aging of tissues, this question is based on the implications of D-Asp accumulation during aging and its implication in the pathogenesis of Alzheimer's disease, arteriosclerosis, and cataracts.

In mammalian tissues the D-Asp oxidase acts on acidic D-amino acids such as D-Asp, D-Glu and NMDA (D'Aniello et al. 1993a). Different researches have supported that D-Asp induced the D-aspartate oxidase enzyme activity (Homma et al. 2007; Nagasaki, 1994; Yamada et al. 1989). Administration of D-Asp to rodents has been reported to increase D-Asp oxidase activity in the liver and kidney (Nagasaki, 1994; Yamada et al. 1989). By the way in the yeast *Cryptococcus humicolus,* D-Asp oxidase activity also was increased when was cultured in medium containing D-Asp as a nitrogen source (Yamada et al. 1996). Furthermore, Homma (2007) mentioned that D-aspartate oxidase mRNA levels also were increased; indicating that transcription of the gene is regulated by D-Asp. Interestingly the induction was specific to D-Asp, whereas L-Asp and D-Glu were ineffective (Takahashi et al.

2004). D-aspartate oxidase expression in mammals is presumed to be regulated in a similar way; however, the details remain to be clarified (Homma, 2007).

3. Concluding Remarks

The described aspects provide a valuable record about importance and functions of D-Asp at central level. Across the time, D-amino acids have happened from not be considered essential for any vital function in higher organisms to prove its vital importance in different biological processes (Spinelli et al. 2006). D-amino acids are a good tool for differentiate between specific and non-specific endogenous processes (Friedman and Levin, 2011). On other hand, D-amino acids can be employed for differentiate disease tissue of healthy tissue, outstanding to potentially better diagnostic that lead to new therapeutic aims (Friedman, 2010; Friedman and Levin, 2011; McConathy and Goodman 2008; Tsukada et al. 2006). Specifically about D-Asp, besides have potential for pharmacologic use, it is involved in cell-to-cell signaling in neuron. It is implicated in the neuronal mechanisms underlying learning and memory (Topo et al. 2010b). What's more, D-Asp is present in the brain of Alzheimer's patients, at significantly lower concentration than in the brain of healthy subjects (D'Aniello et al. 2011; Fisher et al. 1991, 1998; Topo et al. 2009). It`s action like neuromodulatory in glutamatergic synapses might play a beneficial role in conditions related to mammalian brain pathology (Errico et al. 2009), besides in adult neurogenesis (Kim et al. 2010). Nowadays its paper like endogenous neurotransmitter has been extensively supported (D'Aniello et al. 2011).

4. References

[1] Anaya JP, D'Aniello S. Free amino acids in the nervous system of the amphioxus Branchiostoma lanceolatum. A comparative study. *Int J. Biol. Sci.* 2006; 2(2): 87–92.

[2] Aono H, Yuasa S. Evolutionary Significance of the System Utilizing D-Amino acid Enantiomer. *BioSystems.*1997; 9:151-154.

[3] Brown ER., Piscopo S, Chun JT, Francone M, Mirabile I, D'Aniello A. Modulation of an AMPA-like glutamate receptor (SqGluR) gating by L- and D-aspartic acids. *Amino Acids.* 2007; 32: 53–57.

[4] Bruckner H, Westhauser T. Chromatographic determination of L- and D-amino acids in plants. *Amino Acids.* 2003; 24: 43–55.

[5] Burrone L, Di Giovanni M, Di Fiore MM, Baccari GC, Santillo A. Effects of D-aspartate treatment on D-aspartate oxidase, superoxide dismutase, and caspase 3 activities in frog (Rana esculenta) tissues. *Chem Biodiversity.* 2010; 7:1459–1466.

[6] D'Aniello A, Giuditta A. Identification of D-aspartic acid in the brain of Octopus vulgaris. *J Neurochem.* 1977; 29:1053–1057.

[7] D'Aniello A, Giuditta A. 1978. Presence of D-aspartate in squid axoplasm and in other regions of the cephalopod nervous system. *J Neurochem.* 1978; 31:1107–1108.

[8] D'Aniello A, Vetere A, Petrucelli L. Further study on the specificity of D-amino acid oxidase and D-aspartate oxidase and time course for complete oxidation of D-amino acids. *Comp Biochem Physiol.* 1993a; 105: 731–734.

[9] D'Aniello A, Onofrio D, Pischetola G, D'Aniello G, Vetere A, Petrucelli L, Fisher G. Biological role of D-amino acid oxidase and D-aspartate oxidase. Effects of D-amino acids. *J Biol Chem.* 1993b. 268:26941–26949.

[10] D'Aniello A, Nardi G, Vetere A, Ferguson GP. Occurrence of free D-aspartic acid in the circumoesophageal ganglia of Aplysia fasciata. *Life Sci.* 1993c; 52:733–736.

[11] D'Aniello A, Nardi G, De Santis A, Vetere A, Di Cosmo A, Marchelli R, Dossena A, Fisher G. Free l-amino acids and D-aspartate contents in the nervous system of cephalopoda. *Comp Biochem Physiol.* 1995; 112(4):661–666.

[12] D'Aniello A, Di Cosmo A, Di Cristo C, Annunziato L, Petrucelli L, Fisher G. Involvement of D-Aspartic acid in the synthesis of testosterone in rat testes. *Life Sci.* 1996; 59:97–104.

[13] D'Aniello A, Lee JM, Petrucelli L, Di Fiore MM. Regional decrease of free D-aspartate levels in Alzheimer's disease. *Neurosci. Lett.* 1998a; 250:131–134.

[14] D'Aniello A, Di Fiore MM, D'Aniello G, Colin FE, Lewis G, Setchell BP. Secretion of D-aspartic acid by the rat testis and its role in endocrinology of the testis and spermatogenesis. *FEBS Lett* 1998b; 436:23–27.

[15] D'Aniello A, Di Fiore MM, Fisher GH, Milone A, SeleniA, D'Aniello S, Perna A, Ingrosso D. Occurrence of D-Aspartic and N-methyl-D-aspartic acid in rat neuroendocrine tissues and their role in the modulation of luteinizing hormoneand growth hormone release. *FASEB J* 2000; 14:699–714.

[16] D'Aniello A. D-Aspartic acid: An endogenous amino acid with an important neuroendocrine role. *Brain Res Rev.* 2007; 53:215-234.

[17] D'Aniello G, Tolino A, D'Aniello A, Errico F, Fisher GH, DiFiore MM. The Role of D-Aspartic Acid and N-Methyl-D-Aspartic Acid in the Regulation of Prolactin Release. *Neuroendocrinology.* 2000; 141 (10):3862-3870.

[18] D'Aniello G, Ronsini S, Guida F, Spinelli, P, D'Aniello A. Occurrence of D-aspartic acid in humanseminal plasma and spermatozoa: Possible role in reproduction. *Fertility and Sterility.* 2005. 84 (5):1444-1449.

[19] D'Aniello S, Spinelli P, Ferrandino G, Peterson K, Tsesarskia M, Fisher GH, D'Aniello A. Cephalopod vision involves dicarboxylic amino acids: D-aspartate, L-aspartate and L-glutamate. *Biochem J.* 2005; 386:331-340.

[20] D'Aniello S, Garcia-Fernández J. D-Aspartic acid and L-amino acids in the neural system of the amphioxus Branchiostoma lanceolatum. *Amino Acids.* 2007; 32(1):21–26.

[21] D'Aniello S, Somorjai I, Garcia-Fernandez J, Topo E, D'Aniello A. D-Aspartic acid is a novel endogenous neurotransmitter. *FASEB J.* 2007; 25:1014 –1027.

[22] Dunlop DS, Neidle A, McHale D, Dunlop DM, Lajtha A. The presence of free Daspartic acid in rodents and man. *Biochem Biophys Res.* 1986; 141:27–32.

[23] Errico F, Pirro MT, Affuso A, Spinelli P, De Felice M, D'Aniello A, Di Lauro R. A physiological mechanism to regulate D-aspartic acid and NMDA levels in mammals revealed by D-aspartate oxidase deficient mice. *Gene.* 2006; 374:50–57.

[24] Errico F, Nistico R, Palma G, Federici M, Affuso A, Brilli E, Topo E, Centonze D, Bernardi G, Bozzi Y, D'Aniello A, Di Lauro R, Mercuri NB, Usiello A. Increased

levels of D-aspartate in the hippocampus enhance LTP but do not facilitate cognitive flexibility. *Mol Cell Neurosci.* 2008a; 37:236–246.

[25] Errico F, Rossi S, Napolitano F, Catuogno V, Topo E, Fisone G, D'Aniello A, Centonze D, Usiello A. D-aspartate prevents corticostriatal long-term depression and attenuates schizophrenia-like symptoms induced by amphetamine and MK-801. J. *Neurosci.* 2008b; 28:10404–10414.

[26] Errico F, Napolitano F, Nistico R, Centonze D, Usiello A. D-aspartate: an atypical amino acid with neuromodulatory activity in mammals. *Reviews in the Neuroscience.* 2009; 20(5-6):429-40.

[27] Errico F, Nistico R, Napolitano F, Mazzola C, Astone D, Pisapia T, Giustizieri M, D'Aniello A, Mercuri NB, Usiello A. Increased D-aspartate brain content rescues hippocampal age-related synaptic plasticity deterioration of mice. *Neurobiol Aging.* 2010.

[28] Felbeck H, Wiley S. Free D-amino acids in the tissues of marine bivalves. *Biol Bull.* 1987; 173:252-259.

[29] Fisher GH, D'Aniello A, Vetere A, Padula L, Cusano G, Man EH. Free D-aspartate and D-alanine in normal and Alzheimer brain. *Brain Res Bull.* 1991; 26:983–985.

[30] Fisher GH. Appearance of D-amino acids during aging: D-amino acids in tumor proteins. *EXS.* 1998; 85:109–118.

[31] Fisher GH, Lorenzo N, Abe H, Fujita E, Frey WH, Emory WH, Di Fiore MM, D'Aniello A. Free D-and L-amino acids in ventricular cerebrospinal fluid from Alzheimer and normal subjects. *Amino Acids.* 1998; 15:263–269.

[32] Friedman M. Origin, Microbiology, Nutrition, and Pharmacology of D-Amino Acids. *Chemistry & Biodiversity.* 2010; 17:1491–1530.

[33] Friedman M, Lavin CE. Nutritional and medicinal aspects of D-amino acids. *Amino Acids.* 2011. DOI 10.1007/s00726-011-0915.

[34] Fuchs SA, Berger R, Klomp LW, De Koning TJ. D-amino acids in the central nervous system in health and disease. *Mol Genet Metab.* 2005; 85:168–180.

[35] Fujii N, Kaji Y. Racemization of aspartyl residues of proteins in age-related disease. *Seikagaku.* 2008; 80:287–293.

[36] Furuchi T, Homma H. Free D-aspartate in mammals. *Biol Pharm Bull.* 2005; 28:1566–1570.

[37] Greenstein JP, Birnbaum SM, Otey MC. Optical and enzymatic characterization of amino acids. *J Biol Chem.* 1953; 204:307–321.

[38] Hamase K, Morikawa A, Zaitsu K. D-Amino acids in mammals and their diagnostic value. *J Chromatogr B.* 2002; 781:73–91.

[39] Hamase K, Homma H, Takigawa Y, Fukushima T, Santa T, Imai K. Regional distribution and postnatal changes of D-amino acids in rat brain. *Biochim Biophys Acta.* 1997; 1334:214-222.

[40] Hashimoto AT, Nishikawa T, Oka K, Takahashi K. Endogenous D-Serine in Rat Brain: N-Methyl-D-Aspartate Receptor-Related Distribution and Aging. *J Neurochem.* 1993a; 60:783-786.

[41] Hashimoto A, Nishikawa T, Konno R, Niwa A, Yasumura H, Oka T, Takahashi K. Free D-serine, D-aspartate and D-alanine in central nervous system and serum in mutant mice lacking D-amino acid oxidase. *Neuroscience Letters.* 1993b; 152: 33–36.

[42] Hashimoto A, Kumashiro S, Nishikawa T, Oka T, Takahashi K, Mito T, Takashima S, Doi N, Mizutani Y, Kaneco T, Ootomo E. Embryonic development and postnatal changes in free D-aspartate and D-serine in the human prefrontal cortex. *J Neurochem.* 1993c; 61:348–351.

[43] Hashimoto A, Nishikawa T, Oka T, Hayashi T, Takahashi K. Wide spread distribution of free D-aspartate in rat periphery. *FEBS Lett.* 1993d; 331:4–8.

[44] Hashimoto A, Oka T, Nishikawa T. *Eur J Neurosci.* 1995; 7:1657-1663.

[45] Holopainen I, Kontro P. D-aspartate release from cerebellum astrocytes: modulation of the high K induced releaseby neurotransmitter amino acids. *Neuroscience.* 1990; 36:115–112.

[46] Homma H. Biochemistry of D-aspartate in mammalian cells. *Amino Acids.* 2007; 32:3–11.

[47] Huang AS, Beigneux A, Weil ZM, Kim PM, Molliver ME, Blackshaw S, Nelson RJ, Young SG, Snyder SH. D-Aspartate regulates melanocortin formation and function: Behavioral alterations in D-aspartate oxidase-deficient mice. *J Neurosci.* 2006; 26:2814–2819.

[48] Iezhitsa N, Spasov AA, Zhuravleva NV, Sinolitskii MK, Voronin SP. Comparative study of the efficacy of potassium magnesium l-, d- and dl-aspartate stereoisomers in overcoming digoxinand furosemide-induced potassium and magnesium depletion. *Magnesium Res.* 2004; 17:276–292.

[49] Ilisz I, Pataj Z, Aranyi A, Peter A. High-performance liquid chromatography of biologically important, small epimeric peptides and their L,D-amino acid content. *Mini Rev Med Chem.* 2010; 10:287–298.

[50] Ishida AT, Gordon G. D-Aspartate potentiates the effects of L-glutamate on horizontal cells in goldfish retina. *Proc Natl Acad Sci.* 1981; 78:5890–5894.

[51] Kaji Y, Oshika T, Takazawa Y, Fukayama M, Fujii N. Accumulation of D- β-aspartic acid-containing proteins in age related ocular diseases. *Chem Biodiversity.* 2010; 7:1364–1370.

[52] Kanai Y, Hediger MA. Primary structure and functional characterization of a high-affinity glutamate transporter. *Nature* 1992; 360:467–471.

[53] Katane M, Homma H. D-Aspartate- An important bioactive substance in mammals: A review from an analytical and biological point of view. *J Chromatogr B.* 2011. doi:10.1016/j.jchromb. 2011.03.062.

[54] Kim PM, Duan X, Huang AS, Liu CY, Ming GL, Song H, Snyder SH. Aspartate racemase, generating neuronalD-aspartate, regulates adult neurogenesis. *Proc Natl Acad Sci.* 2010; 107:3175–3179.

[55] Kinouchi T, Ishiura S, Mabuchi Y, Rakami-Manaka Y, Nishio H, Nishiuchi Y, Tsunemi M, Takada K, Watanabe M, Ikeda M, Matsui H, Tomioka S, Kawahara, H, Hamamoto T, Suzuki K, Kagawa Y, Mammalian d-aspartyl endopeptidase: A scavenger for noxious racemized proteins in aging. *Biochem Biophys Res Commun.* 2004; 314: 730–736.

[56] Kirschner DL, Green TK. Separation and sensitive detection of D-amino acids in biological matrices. *J Sep Sci.* 2009; 32:2305–2318.

[57] Koyuncuoglu H, Güngör M, Ang O, Inanç D, AngKücüker M, Sagduyu H, Uysal V. Aggravation by morphine and D-aspartic acid of pyelonephritis induced by i.v. inoculation of Staphylococcus aureus in rats. *Infection.* 1988; 16:42-45.

[58] Kumashiro S, Hashimoto A, Nishikawa T. Free D-serine in post-mortem brains and spinal cords of individuals with and without neuropsychiatric diseases. *Brain Res.* 1995; 681:117-125.

[59] Long Z, Homma H, Lee JA, Fukushima T, Santa T, Iwatsubo TR, Yamada R, Imai K. Biosynthesis of D-aspartate in mammalian cells. *FEBS Lett.* 1998; 434(3):231-235.

[60] McConathy J, Goodman MM. Non-natural amino acids for tumor imaging using positron emission tomography and single photon emission computed tomography. *Cancer Metastasis Rev.* 2008; 27:555–573.

[61] Miao H, Rubakhin SV, Scanlan CR, Wang L, Sweedler JV. D-Aspartate as a putative cell-cell signaling molecule in the Aplysia californica central nervous system. *JNeurochem.* 2006; 97:595–606.

[62] Nagasaki H. Gender-related differences of mouse liver D-aspartate oxidase in the activity and response to administration of D-aspartate and peroxisome proliferators. *Int J Biochem.* 1994; 26:415–423.

[63] Nagata Y, Tanaka K, Lida T, Kera Y, Yamada RH, Nakajima Y, Fujiwara T, Fukumori Y, Yamanaka T, Koga Y, Tsuji S, Kawaguchi-Nagata K. Occurrence of D-amino acids in a few archaea and dehydrogenase activities in hyper thermophilic Pyrobaculum islandicum. *Biochim BiophysActa.* 1999; 1435:160–166.

[64] Neidle A, Dunlop DS. Developmental changes of free D-aspartic acid in the chicken embryo and in the neonatal rat. *Life Sci.* 1990; 46:1517-1522.

[65] Nishikawa, T. 2005. Metabolism and Functional Roles of Endogenous D-Serine in Mammalian Brains. *Biol. Pharm. Bull.* 2005; 28:1561-1565.

[66] Pines G, Danbolt NC, Bjoras M, Zhang Y, Bandahan A, Eide L, Koepsell H, Storm-Mathisen J, Seeberg E, Kanner BI. Cloning and expression of a rat brain L-glutamate transporter. *Nature.* 1992; 360:464–467.

[67] Scanlan C, Shi T, Hatcher NG, Rubakhin SS, Sweedler JV. Synthesis, accumulation, and release of D-aspartate in the Aplysia californica. *CNS J Neurochem.* 2010; 115:1234–1244.

[68] Schell MJ, Cooper, OB, Snyder, SH. D-Aspartate localizations imply neuronal and neuroendocrine roles. *Proc Natl Acad Sci.* 1997; 94(5): 2013–2018.

[69] Scolari MJ, Acosta GB. D-serine: a new word in the glutamatergic neuroglial language. *Amino Acids.* 2007; 33(4):563-574.

[70] Spinelli P, Brown E, Ferrandino G, Branno M, Montarolo PG, D'Aniello E, Rastogi RK, D'Aniello B, Baccari G, Fisher G, D'Aniello A. D Aspartic acid in the nervous system of Aplysia limacine: Possible Role in Neurotransmission. *J Cell Physiol.* 2006; 206:672-681.

[71] Takahashi S, Takahashi T, Kera Y, Matsunaga R, Shibuya H, Yamada RH. Cloning and expression in Escherichia coli of the D-aspartate oxidase gene from the yeast Cryptococcus humicola and characterization of the recombinant enzyme. *J Biochem.* 2004; 135: 533–540.

[72] Topo E, Soricelli A, D'Aniello A, Ronsini S, D'Aniello G. The role and molecular mechanism of D-aspartic acid in the release and synthesis of LH and testosterone in humans and rats. *Reproductive Biology and Endocrinology.* 2009; 7:120.

[73] Topo E, Fisher G, Sorricelli A, Errico F, Usiello A, D'Aniello A. Thyroid hormones and D-aspartic acid, D-aspartate oxidase, D-aspartate racemase, H2O2, and ROS in rats and mice. *Chem Biodiversity.* 2010a; 7:1467–1478.

[74] Topo E, Soricelli A, Di Maio A, D'Aniello E, Di Fiore MM, D'Aniello A. Evidence for the involvement of D-aspartic acid in learning and memory of rat. *Amino Acids.* 2010b; 38:1561–1569.

[75] Tsukada H, Sato K, Fukumoto D, Nishiyama S, Harada N, Kakiuchi T. Evaluation of D-isomers of O-11C-methyl tyrosine and O-18F-fluoromethyl tyrosine as tumor-imaging agents in tumor bearing mice: Comparison with L- and D-11C-methionine. *J Nucl Med.* 2006; 47:679–688.

[76] Wang H, Wolosker H, Pevsner J, Snyder S, Selkoe DJ. Regulation of rat magno cellular neurosecretory system by D-aspartate: evidence for biological role(s) of a naturally occurring free D-amino acid in mammals. *J Endocrinol.* 2000; 167(2):247-252.

[77] Wolosker H, D'Aniello A, Snyder SH. D-aspartate disposition in neuronal and endocrine tissues: Ontogeny, biosynthesis and release. *Neuroscience.* 2000; 100:183–189.

[78] Yamada RH, Nagasaki H, Nagata Y, Wakabayashi Y, Iwashima A. Administration of D-aspartate increases D-aspartate oxidase activity in mouse liver. *Biochim Biophys Acta.* 1989; 990:325–328.

[79] Zaar K, Köst HP, Schad A, Völkl A, Baumgart E, FahimiHD. Cellular and subcellular distribution of D-aspartate oxidase in human and rat brain. *J Comp Neurol.* 2002; 450(3):272-282.

[80] Zagon J, Dehnell, Bogl KW. D-Amino acids in organisms and food. *Nutr Res.* 1994; 14:445–463.

In: Tyrosine and Aspartic Acid
Editors: J. E. Jones and D. M. Morano
ISBN: 978-1-62100-752-4

Chapter 7

AMIDE–IMINOL TAUTOMERIZATION OF THE C-TERMINAL PEPTIDE GROUPS OF ASPARTIC ACID RESIDUES. TWO-WATER-ASSISTED MECHANISM, CYCLIZATION FROM THE IMINOL TAUTOMER LEADING TO THE TETRAHEDRAL INTERMEDIATE OF SUCCINIMIDE FORMATION, AND IMPLICATION TO PEPTIDE GROUP HYDROGEN EXCHANGE

***Ohgi Takahashi*[1,*] *and Akifumi Oda*[1,2]**
[1]Faculty of Pharmaceutical Sciences, Tohoku Pharmaceutical University, 4-4-1 Komatsushima, Aoba-ku, Sendai, Miyagi, Japan
[2]Institute for Protein Research, Osaka University, 3-2 Yamadaoka, Suita, Osaka, Japan

ABSTRACT

Amide–iminol tautomerization is an analogue of the well-known keto–enol tautomerization. In this chapter, we show, based on quantum-chemical calculations for a model compound, that iminolization of the C-terminal peptide groups of aspartic acid (Asp) residues may actually occur in peptides and proteins with assistance by two catalytic water molecules. Moreover, it is shown that the iminol tautomer can undergo a two-water-assisted cyclization with relative ease to form a *gem*-diol, which is the precursor of the well-known succinimide imtermediates of isomerization and racemization reactions of Asp residues. These findings indicate that the long-established succinimide-mediated mechanism of the Asp residue reactions may have to be somewhat modified. We also point out that the iminolization mechanism found in the present study may be responsible for the acid-catalyzed hydrogen exchange of the C-terminal peptide groups of Asp residues.

INTRODUCTION

The aspartic acid (Asp) residues (L-Asp residues) in peptides and proteins tend to undergo non-enzymatic isomerization to the β-form (β-Asp) and racemization [1–4]. These spontaneous reactions of Asp residues can alter the three-dimensional structures, properties, and functions of the peptides and proteins. Indeed, these reactions of Asp residues are known to have relevance to age-related diseases such as cataract and Alzheimer's disease [3,4]. Racemization at Asp residues in α-crystallin from aged lenses has been studied in particular detail [3–8]. On the other hand, the Asp residue racemization can be used to date biological materials in archeology, geochemistry, and forensic science, and its clinical applications are also expected [9].

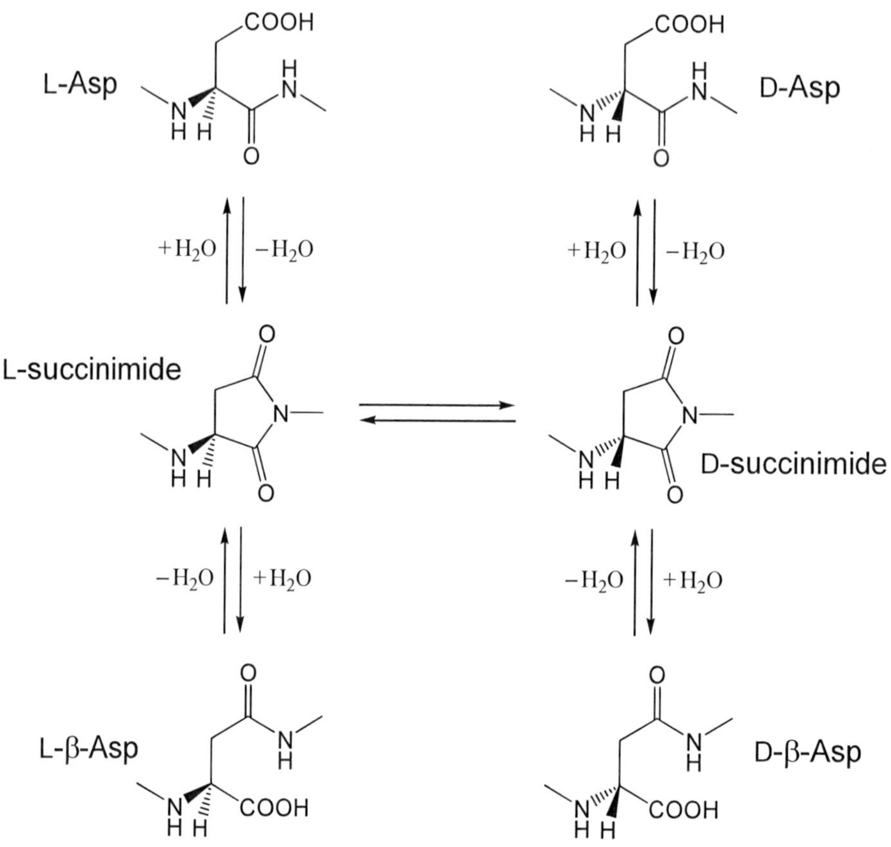

Figure 1. The succinimide-mediated mechanism for the isomerization and racemization reactions of an Asp residue.

For the above spontaneous reactions of Asp residues, the succinimide-mediated mechanism as shown in Figure 1 has been well established [1–4, 9–11]. The Asp residues have propensity to form cyclic succinimide derivatives. The cyclization to give the five-membered succinimide ring results from the nucleophilic attack of the nitrogen atom of the adjacent residue on the C-terminal side, so-called the ($N + 1$) residue, to the carboxyl carbon atom of the Asp residue. Elimination of a water molecule accompanies this cyclization. There are two carbonyl groups in the succinimide moiety, and hydrolysis may occur at either of the two carbonyl groups. Depending on the hydrolysis site, either the L-Asp residue is reproduced or the isomerized L-β-Asp residue is formed. Moreover, the succinimide intermediate is racemization-prone, because its α-hydrogen atom has a relatively high acidity [10,11]. Therefore, its D-form can be formed with relative ease, from which either D-Asp or D-β-Asp residue can be formed. While the succinimide-mediated mechanism is the only one which can explain the formation of the β-Asp residue, a small part of the total racemization may proceed by other mechanisms [1,12].

The immediate product of the nucleophilic attack of the (N + 1) nitrogen atom to the Asp side-chain carboxyl carbon atom is thought to be a *gem*-diol shown in Figure 2 [13–15]. This species is an example of the so-called "tetrahedral intermediates" in nucleophilic substitution reactions. Although such tetrahedral intermediates have been well characterized computationally [13–15], their experimental observation is thought to be difficult. It is required that the side-chain carboxyl group of the Asp residue is in the protonated form (i.e. COOH, not COO^-) in order for its carbon atom to undergo the initial nucleophilic attack. The deprotonated form will not undergo a nucleophilic attack because of its negative charge; moreover, there is no suitable leaving group in the deprotonated form [2]. If we take a pK_a value of 3.9 for the Asp side-chain carboxyl group, only one of every 3000 Asp residues is protonated and is directly available for the nucleophilic reaction at physiological pH of 7.4 [2].

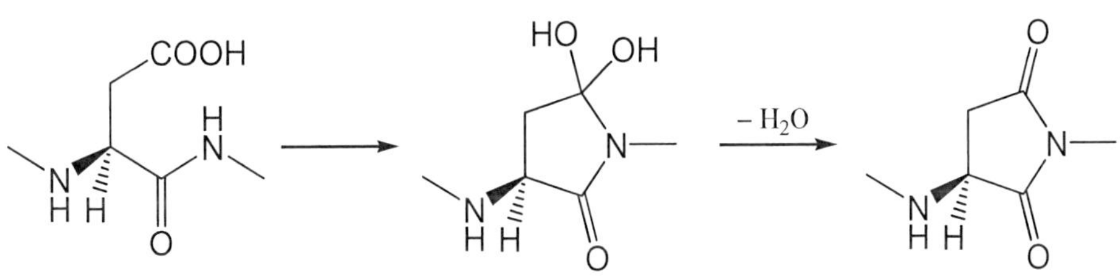

Figure 2. Direct cyclization of an Asp residue leading to the tetrahedral *gem*-diol intermediate and subsequent dehydration to give the succinimide derivative.

Recently, hydrolysis of a succinimide derivative was computationally investigated by Catak et al. in the context of succinimide-mediated deamidation of Asparagine (Asn) residues in peptides and proteins [15]. This hydrolysis reaction is just the reverse reaction of the succinimide formation from the corresponding Asp-containing compound. They used the same methodology as we used in the present study (B3LYP/6-31+G(d,p); see below). Moreover, the hydrolysis product of their calculation was Ace–Asp–Nme, which is the very compound which we used in the present study (see below). They showed that including one explicit water molecule lowers the activation barrier. This water molecule acts as a catalyst by participating in a proton relay. More specifically, according to their results, the succinimide formation from Ace–Asp–Nme proceeds in two steps as follows. In the first step, which is rate-determining, the ring closure (N–C bond formation) and one-water-mediated double hydrogen transfer from the C-terminal NH to the Asp side-chain CO occur concertedly to give a *gem*-diol (so-called tetrahedral intermediate). In the second step, dehydration of the *gem*-diol intermediate occurs assisted by the water molecule. However, the calculated overall activation barrier (corresponding to the first step) was too high (46.1 kcal mol^{-1}) for a biologically relevant reaction. Indeed, Geiger and Clarke reported that the activation energy for the Asp to β-Asp isomerization of a hexapeptide L-Val-L-Tyr-L-Pro-L-Asp-Gly-L-Ala was 21.7 kcal mol^{-1} [1], much lower than the above computational value. Considering that the Asp to β-Asp isomerization can be explained only by intermediacy of the succinimide, it may be that another mechanism has to be invoked to explain the relatively easy formation of succinimide intermediates from Asp residues under physiological conditions.

Here, we note that nucleophilicity (or basicity) of an amide nitrogen atom should not be very high due to the π-conjugation with the C=O group as may be seen from the resonance shown in Figure 3(a). Note that the nitrogen atom has a formal positive charge in the second resonance structure. On the other hand, for the iminol tautomer of an amide, we can write a resonance structure in which the nitrogen atom has a formal negative charge as shown in Figure 3(b). Therefore, the nucleophilicity of an iminol nitrogen atom should be much higher than the nitrogen atom in the corresponding amide form. Therefore, if the C-terminal peptide group of an Asp residue can undergo iminolization with relative ease, then it may be possible that nucleophilic attack of the iminol nitrogen atom to the carboxyl carbon atom occurs to form the cyclic *gem*-diol intermediate (Figure 4).

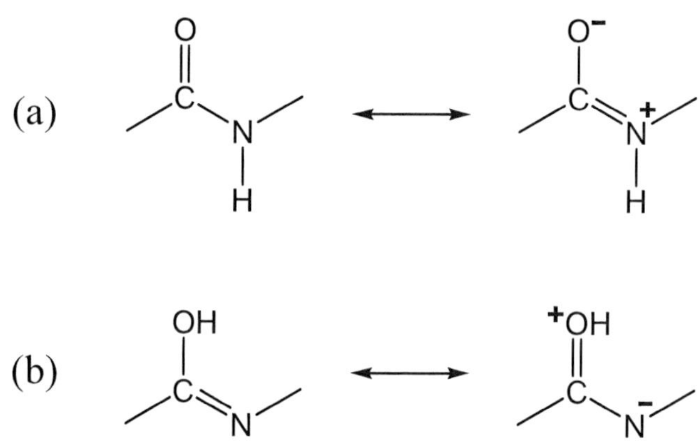

Figure 3. (a) Resonance structures for an amide. (b) Resonance structures for an iminol.

The amide–iminol tautomerization is a nitrogen analogue of the well-known keto–enol tautomerization. The C=C double bond of the enol form of a keto–enol tautomeric pair is replaced by a C=N bond in an amide–iminol tautomerism. There is not much known about the amide–iminol tautomerization of the peptide groups in peptides and proteins [16,17]. In this chapter, we show computationally that the C-terminal peptide group of an Asp residue can easily undergo iminolization by assistance of two catalytic water molecules. Catalysis by two water molecules was recently proposed by us for racemization of the succinimide intermediates [11]. Ace–Asp–Nme was used as the model compound for the present study. Next, we show that, after a conformational change, the iminol tautomer can undergo cyclization to give the *gem*-diol. We also show that this mechanism for the *gem*-diol formation seems to be much preferred to the direct cyclization from the initial amide form. Also, it is pointed out that the iminolization may be partially responsible for the acid-catalyzed hydrogen exchange (HX) of the C-terminal peptide groups of Asp residues.

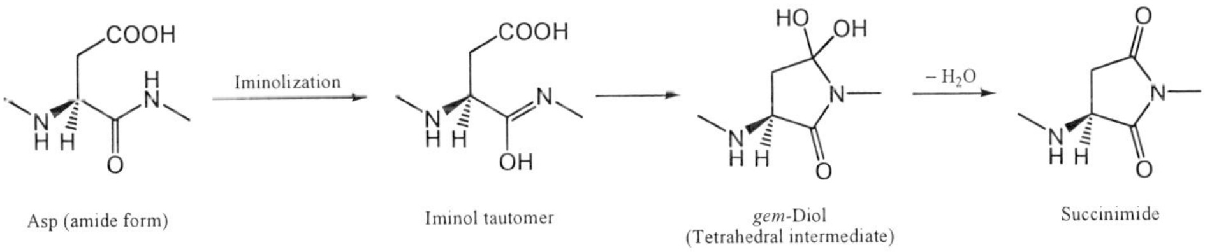

Figure 4. Iminolization of an Asp residue followed by cyclization leading to the tetrahedral *gem*-diol intermediate and subsequent dehydration to give the succinimide derivative.

Computational Method

All calculations were performed using Spartan '08 (Wavefunction, Inc., Irvine, CA, USA). Figure 5 shows the model compound (Ace–Asp–Nme) used in the present study. In this compound, the N- and C-terminals of an Asp residue are capped with Ace and Nme groups, respectively, and there are two amide groups. In the present study, iminolization of the C-terminal amide group was investigated. The side-chain carboxyl group is in the protonated form (i.e. COOH, not COO^-; see above). The complex formed between this model compound and two catalytic water molecules is the reactant in the present computational study. A quantum-chemical method is required to deal with chemical reactions. We employed the density functional theory (DFT) method with the B3LYP functional and the 6-31+G(d,p) basis set. This is the same as used recently by Catak et al. [15] for hydrolysis of the succinimide compound corresponding to the model compound used in the present study.

The geometries of the reactant, iminol intermediates (with different conformations; see below), *gem*-diol product ("tetrahedral intermediate" in the succinimide formation), and transition states (TSs) connecting them were fully optimized by the B3LYP/6-31+G(d,p) method. All geometry optimizations were followed by a vibrational frequency calculation to confirm the nature (i.e., an energy minimum or a TS) of the optimized geometries. The number of imaginary frequencies is zero for energy minima and one for TSs. Intrinsic reaction coordinate (IRC) calculations were performed both in the forward and reverse directions from all the TSs. Successful IRC calculations were followed by a regular geometry optimization. When the IRC calculation was unsuccessful, the atoms were slightly displaced from the TS geometry along the transition vector (i.e., the vibration with an imaginary frequency), followed by a regular geometry optimization. Thus, connectivity of all the calculated energy-minimum and TS geometries was established. The calculated vibrational frequencies were also used to correct the energy of each optimized geometry for the zero-point energy (ZPE).

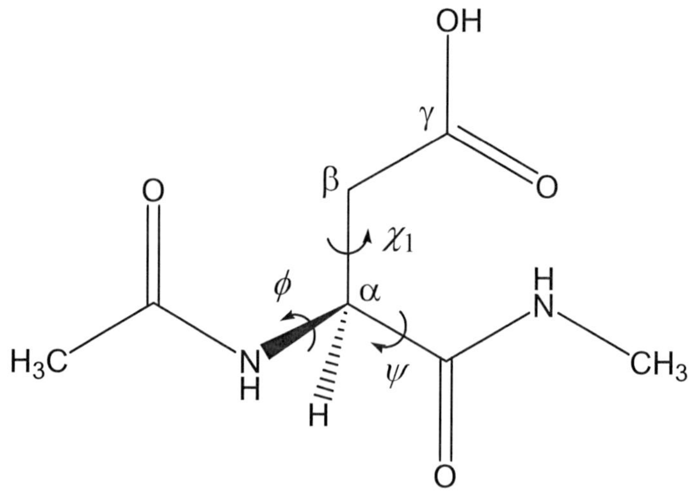

Figure 5. The model compound used in the present study (Ace−Asp−Nme). The Asp residue is in the L-form and its side chain is in the protonated form. The ϕ (C−N−C_α−C) and ψ (N−C_α−C−N) dihedral angles which characterize the main-chain conformation and the χ_1 dihedral angle (N−C_α−C_β−C_γ) which characterizes the side-chain conformation are indicated.

RESULTS AND DISCUSSION

Figure 6 shows an energy diagram based on the present computational results. Beginning from the reactant denoted as *AM* (amide form), there are three steps to reach the cyclic tetrahedral *gem*-diol *TH*. *IM1* and *IM2* are different structures of the iminol form. TS_{AI} stands for the TS which connects *AM* and *IM1*, and TS_{IT} stands for the TS which connects *IM2* and *TH*. *IM-TS* is the TS between *IM1* and *IM2*. In the following, these seven key structures are described in detail.

Two-Water-Assisted Iminolization of the Amide Form

The optimized geometry of *AM* is shown in Figure 7. *AM* is the reactant complex formed by the model compound (amide form shown in Figure 5) and two water molecules. The ϕ and ψ dihedral angles which characterize the conformation of the main chain are −103° and −96°, respectively. As to the conformation of the side chain, the χ_1 dihedral angle (defined by N, C_α, C_β, and C_γ) is 167°. Of course, there are many other conformers possible, but *AM* leads directly to TS_{AI}, which is the TS with the lowest energy among those found for the iminolization with two catalytic water molecules. The (ϕ, ψ) = (−103°, −96°) conformation is expected to be assumed in short peptides and in structurally flexible regions of proteins. Furthermore, this conformation may even be achieved from an α-helix structure (at least after partial denaturation). The ideal ϕ and ψ values for an L-amino acid in an α-helix are about −57° and −47°, respectively. Indeed, it was recently shown that Asp residues in an α-helix structure of (Leu-Asp-Asp-Leu)$_8$-Asp racemize more rapidly than those in a β-sheet structure of (Asp-Leu)$_{15}$ [18]. In parallel and antiparallel β-sheet structures, the dihedral angle ψ takes positive values of about 115° and 135°, respectively. Therefore, a conformation near ψ = −96° will be quite difficult to be reached from a β-sheet structure.

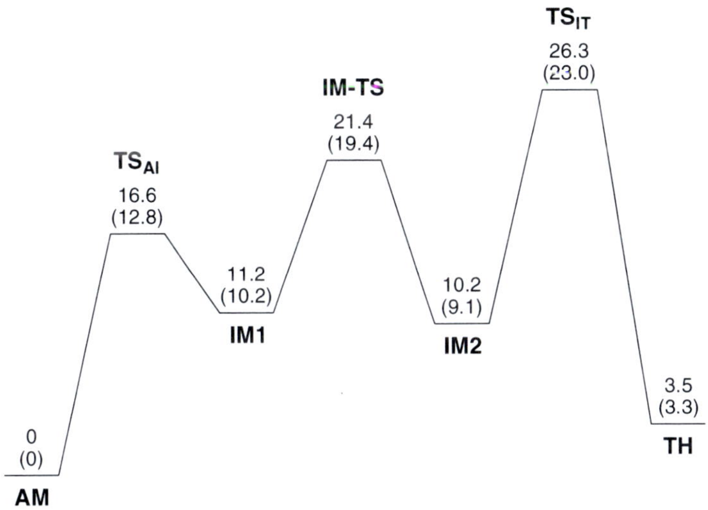

Figure 6. Energy diagram for the two-water-assisted reaction of the model compound shown in Figure 5 to form the *gem*-diol species. For abbreviations, see text. Relative energies with respect to *AM* before and after (in parentheses) ZPE corrections are shown in kcal mol^{-1}.

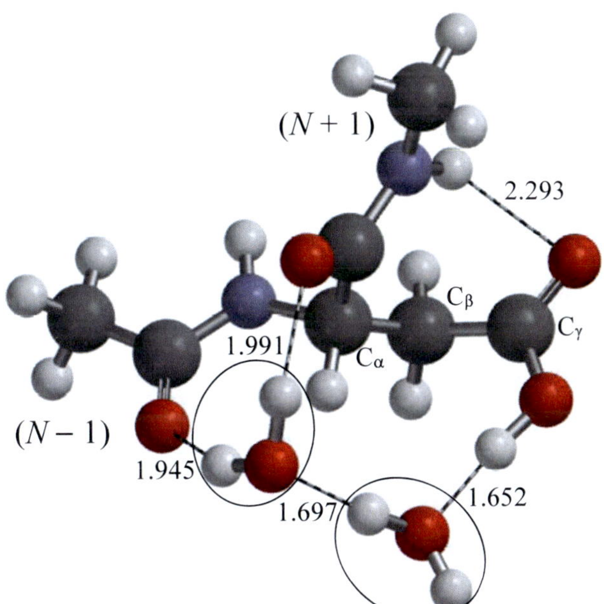

Figure 7. The optimized geometry of *AM* (the two catalytic water molecules are circled). Hydrogen bond distances are shown in Å. The ϕ, ψ, and χ_1 dihedral angles are −103°, −96°, and 167°, respectively. The total energy of this geometry is −837.373515 E_h.

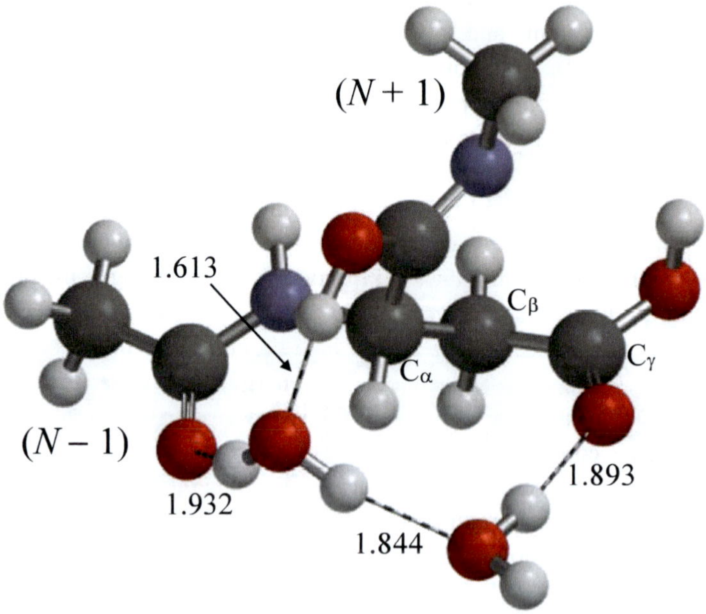

Figure 10. The optimized geometry of *IM-TS*. Selected interatomic distances are shown in Å. The ϕ, ψ, and χ_1 dihedral angles are −113°, −104°, and 175°, respectively. The imaginary frequency of this TS is 578i cm^{-1}.

Figure 9 shows the geometry of *IM1*, which is formed from *AM* in one step through TS_{AI} just described. The values of the ϕ, ψ, and χ_1 dihedral angles in *IM1* are −111°, −100°, and −176°, respectively, and are close to the corresponding values in TS_{AI}. There is a hydrogen bond (1.839 Å) between the iminol nitrogen atom and the newly formed OH group of the side chain. This OH group is *trans* to the newly formed CO group. The hydrogen bond between the iminol OH group and one of the water molecules is very short (1.618 Å). This water molecule also forms a hydrogen bond (1.936 Å) to the (N − 1) CO as in *AM* and TS_{AI}. The relative energies of *IM1* before and after the ZPE correction are 11.2 and 10.2 kcal mol^{-1}, respectively, with respect to *AM*. Since the energy barrier to the reaction back to *AM* is as low as 2.6 kcal mol^{-1}, it would be difficult to directly observe an iminol tautomer experimentally.

Structural change in the Iminol Tautomer

We could not found a pathway in which direct cyclization from *IM1* to the *gem*-diol occurs. It does not seem that *IM1* is "ready" to cyclize, because, as described above, the nitrogen atom of the iminol moiety, which is supposed to attack the carboxyl carbon atom, forms a hydrogen bond to the OH group of the side chain. Indeed, in order to reach TS_{IT}, which is the lowest-energy TS for cyclization among those we found, *IM1* has to undergo a structural change to *IM2*. This change mainly consists of the rotation of the side-chain OH group and the rotation about the C_β–C_γ single bond.

IM-TS is the TS for this structural change and Figure 10 shows its geometry. The ϕ, ψ, and χ_1 dihedral angles are −113°, −104°, and 175°, respectively, in *IM-TS*, and again close to the respective former values in *AM*, TS_{AI}, and *IM1*. The C–O bond in the side-chain carboxyl group is almost perpendicularly twisted, and the hydrogen bond between the iminol nitrogen and the side-chain OH has been broken (the N•••H distance is 2.806 Å). Also, considerable rotation about the C_β–C_γ bond has occurred; the C_α–C_β–C_γ=O dihedral angle was 92° in *IM1*, but is 46° in *IM-TS*. As in *IM1*, the hydrogen bond between the iminol OH and one of the water molecules is very short (1.613 Å). The hydrogen bond between this water molecule and the (*N* – 1) CO is 1.932 Å.

The ZPE-corrected energy of *IM-TS* is higher than that of *IM1* by about 9 kcal mol^{-1}. *IM-TS* is even higher in energy than TS_{AI}. However, as mentioned below, the energy of TS_{IT} (the TS of the next step) is even higher, and the structural change from *IM1* to *IM2* is not the rate-determining step of the overall reaction.

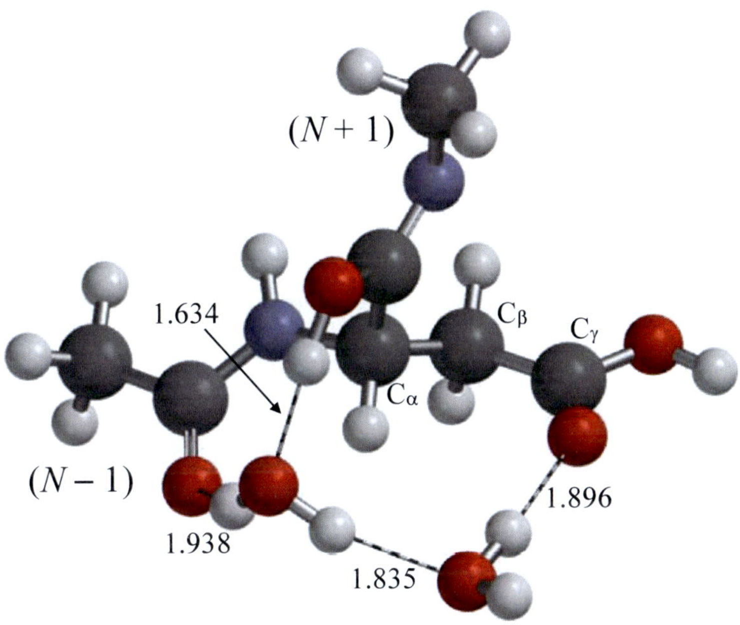

Figure 11. The optimized geometry of *IM2*. Hydrogen bond distances are shown in Å. The ϕ, ψ, and χ_1 dihedral angles are –114°, –93°, and –172°, respectively.

Figure 11 shows the geometry of *IM2*. With respect to the side-chain CO group, the side-chain OH group was *trans* in *IM1*, but is *cis* in *IM2*. The C_α–C_β–C_γ=O dihedral angle has been reduced to 23°. As a result, the iminol nitrogen atom does not have a hydrogen bond and is "ready" to attack the carboxyl carbon atom, although the distance between the nitrogen and carbon atoms is not very short (3.262 Å). Actually, this distance is slightly longer than that in *IM1* (3.143 Å). However, there is nothing to disturb the bond formation between the nitrogen and carbon atoms in *IM2*. Interestingly, the N•••C distance was considerably shorter in *IM-TS* (2.853 Å).

Again, the iminol OH forms a short hydrogen bond (1.634 Å) to the water molecule which also forms a hydrogen bond (1.938 Å) to the (*N* – 1) CO. *IM2* is more stable than *IM1* by about 1 kcal mol^{-1}. Therefore, the conversion of *IM1* to *IM2* is energetically favored. The ϕ, ψ, and χ_1 values of *IM2* are –114°, –93°, and –172°, respectively. Thus, the overall change in shape is relatively small throughout the conversion of *AM* to *IM2*, as may be seen by comparing Figures 7–11.

Cyclization from the Iminol Tautomer to Form the Cyclic Tetrahedral Intermediate

The nucleophilic attack of the iminol nitrogen atom of *IM2* to the side-chain carboxyl carbon atom results in cyclization to form a five-membered ring. A triple hydrogen transfer from the iminol OH group to the side-chain CO group by participation of the two water molecules and π-bond reorganization of the OCN moiety couple with the cyclization (i.e., bond formation between the nitrogen and carbon atoms), so that the product of this single step is the tetrahedral *gem*-diol *TH*. The corresponding TS is TS_{IT}, whose geometry is shown in Figure 12. As noted above, TS_{IT} is higher in energy than those of the two former TSs (i.e., TS_{AI} and *IM-TS*). Therefore, the cyclization of *IM2* is the rate-determining step for the overall reaction shown in Figure 6. The relative energy of TS_{IT} with respect to the initial reactant complex *AM* is 26.3 kcal mol^{-1} before the ZPE correction; this is lowered to 23.0 kcal mol^{-1} by the ZPE correction. The latter value, which represents the overall activation barrier, is in reasonable agreement with the experimental activation energy of 21.7 kcal mol^{-1} determined for the Asp to β-Asp isomerization of the hexapeptide L-Val-L-Tyr-L-Pro-L-Asp-Gly-L-Ala by the Arrhenius plot [1]. This isomerization occurs through the succinimide intermediate which is thought to be formed from the tetrahedral *gem*-diol precursor by dehydration. Therefore, if the energy of the TS for the dehydration is lower than or comparable to that for the *gem*-diol-forming cyclization TS, it is highly probable that the iminolization precedes cyclization. Although computational study on the dehydration of *TH* is now in progress, preliminary results indicate that the TSs for the dehydration have comparable energies to TS_{IT}. Therefore, it may be highly possible that the iminolization–cyclization–dehydration mechanism is responsible for the most of succinimide formation from Asp residues of peptides and proteins in neutral aqueous media. Thus, the long-established mechanism of the succinimide formation (i.e., cyclization–dehydration) may have to be modified by including the iminolization step.

The ϕ, ψ, and χ_1 dihedral angles of TS_{IT} are −96°, −136°, and 157°, respectively. The changes in these dihedral angles on going from *IM2* to TS_{IT} are larger than those observed between *AM* and *IM2*. Nevertheless, the overall change in shape observed between *IM2* and TS_{IT} is not drastic but rather smooth. This is important for the cyclization reactions in proteins, in which there are structural constraints imposed by their three-dimensional structures. The distance of the forming N–C bond in TS_{IT} is 1.621 Å, and formation of this bond is synchronized with the other major structural change, i.e., hydrogen transfers and π-bond reorganization. The hydrogen bond of the $(N - 1)$ CO is rather short (1.736 Å) in TS_{IT}.

The geometry of *TH* is shown in Figure 13. The three dihedral angles, ϕ, ψ, and χ_1, are −98°, −141°, and 152°, respectively. These values are very close to the corresponding values in TS_{IT} through which *TH* is formed. *TH* is higher in energy than *AM* by about 3 kcal mol^{-1}, and the energy minimum for *TH* seems to be rather deep. It may be that the tetrahedral intermediates of succinimide-forming reactions are more long-lived than have been postulated. If so, there is possibility that a small part of Asp residue racemizations is due to the racemization at the stage of the tetrahedral intermediate. Indeed, it was pointed out by

Geiger and Clarke [1] that more than two thirds of the total Asp racemization observed for the hexapeptide L-Val-L-Tyr-L-Pro-L-Asp-Gly-L-Ala can be attributed to the succinimide intermediate but non-succinimide-mediated mechanisms have to be invoked for the rest. Moreover, it was experimentally shown that, in a basic condition, racemization of the asparagine (Asn) residues in the pentapeptides Gly-L-Gln-(L,D)-Asn-L-Glu-Gly occured at the stage of the tetrahedral intermediate similar to *TH* in the course of the succinimide-mediated deamidation of these residues [19]. Therefore, the possibility of racemization at the stage of the tetrahedral intermediate like *TH* can not be eliminated [20,21].

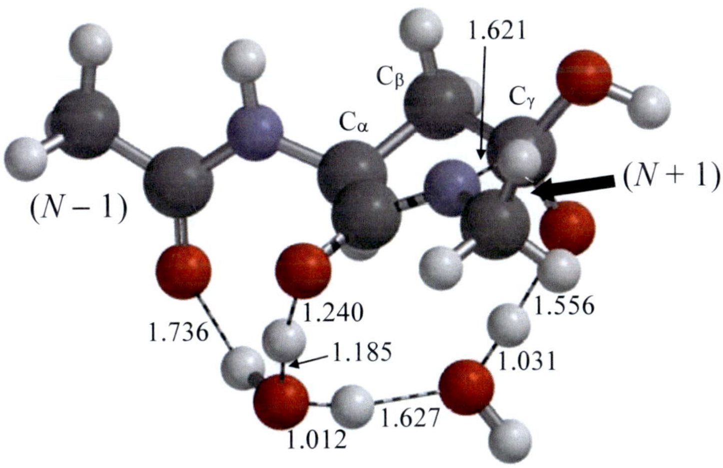

Figure 12. The optimized geometry of TS_{IT}. Selected interatomic distances are shown in Å. The ϕ, ψ, and χ_1 dihedral angles are −96°, −136°, and 157°, respectively. The imaginary frequency of this TS is 428i cm^{-1}.

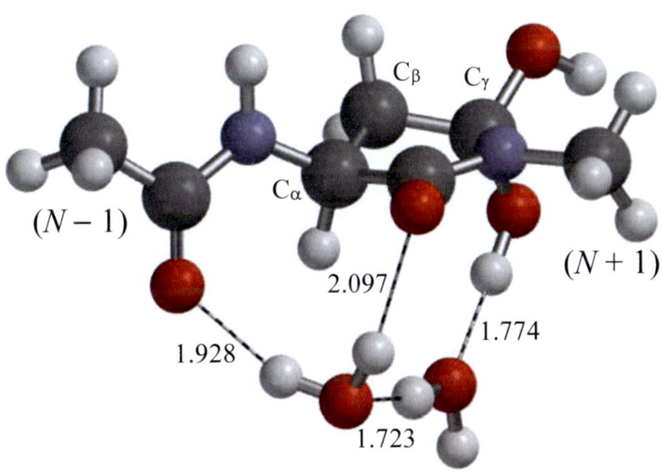

Figure 13. The optimized geometry of **TH**. Hydrogen bond distances are shown in Å. The ϕ, ψ, and χ_1 dihedral angles are −98°, −141°, and 152°, respectively.

Comparison with the Direct Cyclization from the Amide Form

In order to further support the iminol-mediated mechanism proposed above for the succinimide formation from Asp residues, it is required to show that direct cyclization of the Asp residue to give the *gem*-diol has a higher activation energy. For comparison purposes, use of the same number of catalytic water molecules (i.e., two catalytic water molecules) is desirable. Therefore, we performed TS searches for the direct cyclization of the model compound (Figure 5) complexed with two water molecules by the same method (B3LYP/6-31+G(d,p)) as used for the above calculations. In this case, the two water molecules make a bridge between the ($N + 1$) NH and the side-chain CO.

We performed extensive TS searches and found several TSs with different conformations which connect the amide form and the *gem*-diol species. However, all of them had much higher energies (by > 9 kcal mol^{-1}) than TS_{IT}, the TS of the rate-determining step in the iminol-mediated mechanism. Moreover, we could not found a direct cyclization TS which has a similar main-chain and side-chain conformation to TS_{IT}, the iminol cyclization TS. Figure 14 shows one of the TSs found for the direct cyclization. Its energy is higher than that of TS_{IT} by 12.5 and 9.1 kcal mol^{-1} before and after the ZPE correction, respectively. Although the main-chain conformation of this TS ($\phi = -119°$, $\psi = -127°$) is similar to TS_{IT}, the side-chain conformation is not ($\chi_1 = 96°$).

Moreover, the reactant (amide form) and the product (*gem*-diol) connected by this TS are not *AM* (Figure 7) and *TH* (Figure 13), respectively, but have (ϕ, ψ, χ_1) = (−166°, 168°, 74°) and (−121°, −100°, 92°) conformations, respectively (geometries not shown). These conformations are largely different from those of *AM* and *TH*, respectively.

Because we could not found a direct cyclization TS with an energy lower than or comparable to TS_{IT}, we can not expect that the direct cyclization is the major mechanism for the *gem*-diol formation. The nucleophilic attack by the ($N + 1$) nitrogen atom is predicted to occur after iminolization as we anticipated (see INTRODUCTION). This means a modification of the long-established succinimide-mediated mechanism of Asp residue reactions.

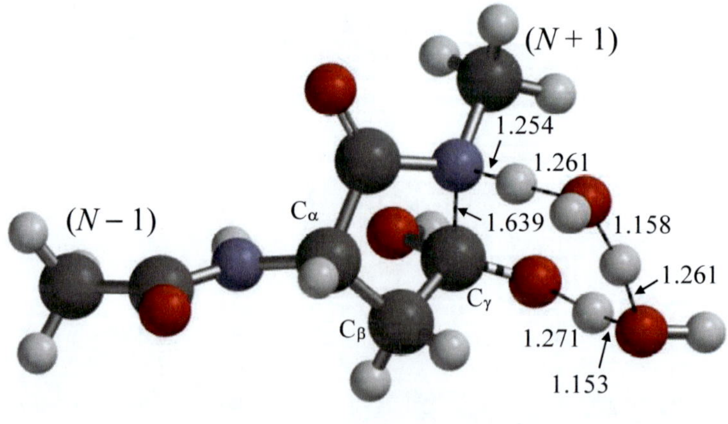

Figure 14. The geometry of a TS found for the direct cyclization (with two catalytic water molecules) of the model compound. Selected interatomic distances are shown in Å. The ϕ, ψ, and χ_1 dihedral angles are −119°, −127°, and 96°, respectively. The imaginary frequency of this TS is 1352i cm^{-1}.

Implication to the Mechanism of Acid-Catalyzed Hydrogen Exchange of the C-Terminal Peptide Groups of Asp Residues

In the above, it has been predicted that iminolization of the C-terminal peptide groups of Asp residues can actually occur in neutral aqueous media. Since the hydrogen atom attached to the peptide nitrogen atom is abstracted, this reaction may cause hydrogen exchange (HX) of the peptide group. In general, peptide group HX behavior of peptides and proteins contains information about their structures, conformational dynamics, and interactions, and acid- and base-catalyzed HX reactions have been extensively studied for small model compounds [22,23].

For acid-catalyzed HX reactions of peptide groups, the "imidic acid mechanism" has been accepted for a long time [24]. The "imidic acid" is what we call "iminol" in this chapter. In this mechanism, protonation at the peptide oxygen atom occurs first, and then deprotonation from the peptide nitrogen atom occurs to give an iminol intermediate, via which the NH hydrogen is exchanged.

For the acid-catalyzed HX of the C-terminal peptide group of an Asp residue, both the protonated (COOH) and deprotonated (COO^-) forms of the side chain have to be considered, since the pK_a value for the side-chain carboxyl group is about 3.9. It was experimentally shown for Ace–Asp–Nme, which is just the model compound itself used in the presnt study, that acid-catalyzed HX of the C-terminal peptide group from the deprotonated form is accelerated compared to the Alanine (Ala) analogue [22,23]. In this case, however, protonation should occur at the side-chain COO^- group, not at the main-chain peptide oxygen atom. Therefore, the iminol mechanism has to be modified for this case.

As shown above, two-water-assisted iminolization of the C-terminal peptide group of Ace–Asp–Nme (protonated form) is expected to occur easily. In this reaction, the hydrogen atom attached to the peptide nitrogen atom is abstracted by the side-chain CO oxygen atom. Exchange of this hydrogen atom with another one in the solvent water molecules and subsequent return to the amide form can result in HX. Therefore, the two-water-assisted iminolization may be partially involved in the acid-catalyzed HX of the C-terminal peptide groups of Ace–Asp–Nme and other Asp residues in peptides and proteins.

CONCLUSION

DFT quantum-chemical calculations on the Asp-containing model compound Ace–Asp–Nme provided several new insights into the chemistry of Asp residues. First, iminolization of the C-terminal peptide groups of Asp residues can occur with relative ease from their protonated forms, if conformational requirements are fulfilled and two water molecules are available; these water molecules act as catalyst by participating in a proton relay. A value of about −100° for both the ϕ and ψ dihedral angles was predicted to be most suitable for the iminolization. Second, cyclization to give the tetrahedral *gem*-diol species can occur from a conformer of the iminol tautomer. This step is also catalyzed by two water molecules which participate in a proton relay. The *gem*-diol is the precursor of the succinimide intermediate in the isomerization and racemization reactions of Asp residues. Third, the iminolization–cyclization mechanism for the *gem*-diol formation seems to be preferred over the direct cyclization from the initial amide form of the Asp residue. Therefore, the well-known succinimide-mediated mechanism for the Asp residue reactions may have to be modified by including the iminolization step. Finally, the two-water-assisted iminolization of the C-terminal peptide groups of Asp residues may be partially responsible for the hydrogen exchange of these peptide groups. Computational studies to further support the involvement of the amide–iminol tautomerization in the chemistry of Asp residues are now in progress.

REFERENCES

[1] Geiger, T. & Clarke, S. (1987). Deamidation, isomerization, and racemization at asparaginyl and aspartyl residues in peptides. Succinimide-linked reactions that contribute to protein degradation. *J Biol Chem*, *262*, 785–794.

[2] Stephenson, R. C. & Clarke, S. (1989). Succinimide formation from aspartyl and asparaginyl peptides as a model for the spontaneous degradation of proteins. *J Biol Chem*, *264*, 6164–6170.

[3] Ritz-Timme, S. & Collins, M. J. (2002). Racemization of aspartic acid in human proteins. *Aging Res Rev*, *1*, 43–59.

[4] Fujii, N. (2005). D-Amino acid in elderly tissues. *Biol Pharm Bull*, *28*, 1585–1589.

[5] Fujii, N., Satoh, K., Harada, K. & Ishibashi, Y. (1994). Simultaneous stereoinversion and isomerization at specific aspartic acid residues in αA-crystallin from human lens. *J Biochem*, *116*, 663–669.

[6] Fujii, N., Ishibashi, Y., Satoh, K., Fujino, M. & Harada, K. (1994). Simultaneous racemization and isomerization at specific aspartic acid residues in αB-crystallin from the aged human lens. *Biochim Biophys Acta*, *1204*, 157–163.

[7] Fujii, N., Takemoto, L. J., Momose, Y., Matsumoto, S., Hiroki, K. & Akaboshi, M. (1999). Formation of four isomers at the Asp-151 residue of aged human αA-crystallin by natural aging. *Biochem Biophys Res Commun*, *265*, 746–751.

[8] Fujii, N., Matsumoto, S., Hiroki, K. & Takemoto, L. (2001). Inversion and isomerization of Asp-58 residue in human αA-crystallin from normal aged lenses and cataractous lenses. *Biochim Biophys Acta*, *1549*, 179–187.

[9] McCudden, C. R. & Kraus, V. B. (2006). Biochemistry of amino acid racemization and clinical application to musculoskeletal disease. *Clin Biochem*, *39*, 1112–1130.

[10] Radkiewicz, J. L., Zipse, H., Clarke, S. & Houk, K. N. (1996). Accelerated racemization of aspartic acid and asparagine residues via succinimide intermediates: An ab initio theoretical exploration of mechanism. *J Am Chem Soc*, *118*, 9148–9155.

[11] Takahashi, O., Kobayashi, K. & Oda, A. (2010). Modeling the enolization of succinimide derivatives, a key step of racemization of aspartic acid residues: Importance of a two-H_2O mechanism. *Chem Boidiv*, *7*, 1349–1356.

[12] Takahashi, O., Kobayashi, K. & Oda, A. (2010). Computational modeling of the enolization in a direct mechanism of racemization of the aspartic acid residue. *Chem Boidiv*, *7*, 1630–1633.

[13] Konuklar, F. A., Aviyente, V. & Ruiz Lopez, M. F. (2002). Theoretical study on the alkaline and neutral hydrolysis of succinimide derivatives in deamidation reactions. *J Phys Chem A*, *106*, 11205–11214.

[14] Aylin, F., Konuklar, S. & Aviyente, V. (2003). Modelling the hydrolysis of succinimide: Formation of aspartate and reversible isomerization of aspartic acid via succinimide. *Org Biomol Chem*, *1*, 2290–2297.

[15] Catak, S., Monard, G., Aviyente, V. & Ruiz-López, M. F. (2009) Deamidation of asparagine residues: Direct hydrolysis versus succinimide-mediated deamidaion mechanisms. *J Phys Chem A*, *113*, 1111–1120.

[16] Kamiya, K., Boero, M., Shiraishi, K. & Oshiyama, A. (2006). Enol-to-keto tautomerism of peptide groups. *J Phys Chem B*, *110*, 4443–4450.

[17] Li, Q., Xue, Y. & Yan, G. (2008). Water-assisted enol-to-keto tautomerism of a simple peptide model: A computational investigation. *J Mol Struct: THEOCHEM*, *868*, 55–64.

[18] Kuge, K., Brack, A. & Fujii, N. (2007). Conformation-dependent racemization of aspartyl residues in peptides. *Chem Eur J*, *13*, 5617–5621.

[19] Li, B., Borchardt, R. T., Topp, E. M., VanderVelde, D. & Schowen, R. L. (2003). Racemization of an asparagine residue during peptide deamidation. *J Am Chem Soc*, *125*, 11486–11487.

[20] Conrad, U., Taichrib, A., Neusüß, C. & Scriba, G. K. E. (2010). Capillary electrophoresis analysis of the degradation of the aspartyl tripeptide Phe-Asp-GlyOH at pH 2.0 and 7.4 under forced conditions. *J Pharm Biomed Anal*, *51*, 640–648.

[21] Conrad, U., Fahr, A. & Scriba, G. K. E. (2010). Kinetics of aspartic acid isomerization and enantiomerization in model aspartyl tripeptides under forced conditions. *J Pharm Sci*, *99*, 4162–4173.

[22] Molday, R. S., Englander, S. W. & Kallen, R. G. (1972). Primary structure effects on peptide group hydrogen exchange. *Biochem*, *11*, 150–158.

[23] Bai, Y., Milne, J. S., Mayne, L. & Englander, S. W. (1993). Primary structure effects on peptide group hydrogen exchange. *Proteins Struct. Funct. Genet.*, *17*, 75–86.

[24] Perrin, C. L., Lollo, C. P. & Johnston, E. R. (1984). NMR site-to-site rate constants and the mechanisms of acid-catalyzed proton exchange in secondary amides. *J Am Chem Soc*, *106*, 2749–2753.

In: Tyrosine and Aspartic Acid
Editors: J. E. Jones and D. M. Morano
ISBN: 978-1-62100-752-4

Chapter 8

Physiological Role of RTK Transactivation in the Central Nervous System and Neuropsychiatric Diseases

Eiki Takahashi*

Research Resources Center, RIKEN Brain Science Institute, Hirosawa 2-1, Wako, Saitama, Japan

Abstract

G protein-coupled receptors (GPCRs) generate signals that promote gene transcription through the 'transactivation' of receptor tyrosine kinases (RTKs) such as the receptors for epidermal growth factor, fibroblast growth factor, platelet-derived growth factor (PDGF), and neurotrophins. RTKs and downstream signaling pathways activated in response to GPCR-mediated stimulation elicit phenotypic responses. Transactivation is essential for the specificity and fidelity of signal transduction and biological responses and is particularly important in the central nervous system (CNS), where RTKs mediate key events in the development and maintenance of neuronal circuits. Recent evidence indicates that RTK transactivation may be involved in the pathophysiology of neuropsychiatric diseases. Dopamine, a major transmitter and neuromodulator in the CNS, is implicated in psychiatric diseases. A GPCR, the D4 dopamine receptor, inhibits 1) excitatory transmission mediated by the N-methyl-D-aspartate (NMDA) glutamate receptor subtype and 2) synaptic transmission through transactivation of the PDGF receptor. The demonstration of a physiological role for RTK transactivation in the CNS provides opportunities for understanding how aberrant dopamine signaling might contribute to neuropsychiatric diseases and for producing novel treatment strategies for these diseases.

Keywords: G-protein-coupled receptor (GPCR); neuropsychiatric disease; receptor tyrosine kinase (RTK); signaling pathway; transactivation

* Corresponding author: Eiki Takahashi, DVM, PhD, DJCLAM. Research Resources Center, RIKEN Brain Science Institute, Hirosawa 2-1, Wako, Saitama, Japan, 351-0198, E-mail: etakahashi@brain, Telephone: 81-48-467-5871, Facsimile: 81-48-467-9692.

Introduction

Agonist-stimulated G protein-coupled receptors (GPCRs) cause activation of heterotrimeric G proteins by catalyzing the exchange of GDP for GTP on the Gα subunit and dissociation of GTP-bound Gα from Gα–Gβγ heterodimers. The dissociated Gα and Gβγ subunits interact with a variety of molecules, including adenylyl cyclase, phospholipase C (PLC) isoforms, and ion channels. The interactions of Gsα and Giα with adenylyl cyclases respectively stimulate and inhibit the production of intracellular cAMP. Activation of PLCβ by Gqα promotes the formation of inositol (1,4,5)-trisphosphate (IP3) and diacylglycerol (DAG), leading to elevation of intracellular Ca^{2+} levels and activation of protein kinase C (PKC) isoforms [51, 58]. The second messengers cAMP, Ca^{2+}, and PKC, and other signaling intermediates such as Src, Pyk2, reactive oxygen species (ROS), and phosphatidylinositol 3-kinase (PI3-K) contribute to the regulation of neuronal proliferation, survival, development, differentiation, migration, long-term potentiation (LTP), and synaptic transmission in the central nervous system (CNS) [9, 16, 78].

Receptor tyrosine kinases (RTKs) belong to a broad family of transmembrane receptors that possess intrinsic kinase activity. RTKs are essential for development and maintenance of the nervous system and regulate diverse neural processes, including cell survival, proliferation, differentiation, neurite outgrowth, and synaptogenesis [45, 82]. Activation of RTKs by their cognate growth factor or polypeptide ligands results in the autophosphorylation of specific tyrosine residues within the C-terminal tyrosine kinase domain of the receptor. This leads to receptor interactions with Src homology-2 domain-containing adaptor and signaling proteins and the activation of downstream signaling pathways including the extracellular-regulated kinase (ERK) mitogen-activated protein kinase (MAPK) pathway, as well as the PI-3K, Akt, and PLC-γ pathways [29, 56].

It has become evident that GPCRs and RTKs, which activate a common set of signaling molecules, do not operate in an isolated fashion. It is well established that the growth-promoting activity of many GPCR ligands involves activation of RTKs and their downstream signaling cascades. For example, both GPCRs and RTKs generate signals that control neuronal cells, some of which involve stimulation of the Ras family of GTPases or activation of MAPK signaling [68, 79, 80]. These observations have led to the emergence of the 'transactivation' concept, according to which a given receptor is activated by the ligand for a heterologous receptor, which possibly belongs to class of receptors with a different signal transduction mechanism.

Mounting recent evidence indicates that RTK transactivation may be involved in the pathophysiology of neuropsychiatric diseases. This review summarizes what is known about the physiological role of RTK transactivation in the CNS and discusses how aberrant dopamine signaling contributes to neuropsychiatric diseases and novel strategies for treating them.

RTK Signaling in Neurons

Ligand binding-induced activation and autophosphorylation results in recruitment of several adapter molecules to tyrosine residues in the intracellular domains of RTKs. Adapter proteins then trigger the activation of several intracellular signaling pathways. The main cascades activated by RTKs include the Ras/Raf/MEK/ERK, PI3-K/Akt, and PLC pathways. ERK and Akt signaling is involved in both neuronal survival and axon growth, and PLC mainly regulates intracellular Ca^{2+}-levels and PKC activity via cleavage of phosphatidylinositol 4,5-bisphosphate (PIP2) to DAG and IP3.

Transactivation of the EGF Receptor by GPCRs

Members of the epidermal growth factor (EGF) receptor family, which include ErbB1, ErbB2, ErbB3, and ErbB4, are present in the CNS and mediate the actions of a family of growth factors including EGF, transforming growth factor (TGF)-α, and neuregulins [8, 43]. The growth-promoting effects of GPCRs are mediated through transactivation of the EGF receptor, and they mainly elicit cellular responses via their intrinsic tyrosine kinase activity. Blocking EGF receptor kinase activity with the highly selective antagonist AG1478, antisense oligonucleotides, or a kinase-negative EGF receptor abolishes downstream signaling events such as the activation of ERK1/2 and inhibits proliferation in several cell types [61, 83]. Hypothalamic astroglial ErbB receptors contribute to the initiation of mammalian puberty by promoting the secretion of gonadotropin-releasing hormone (GnRH), the neuropeptide that controls reproductive function [60]. GnRH activates pro-survival signaling events such as the activation of PI3-K and Akt and the phosphorylation of p90 ribosomal protein S6 kinase 1 (RSK-1) and cAMP-response-element-binding protein (CREB) in hypothalamic GT1-7 neuronal cells. These responses result from transactivation of the EGF receptor by GnRH. This process occurs through sequential activation of PKCα/e and Src/Pyk2 and the induction of metalloprotease-dependent ectodomain shedding of soluble ligands such as heparin-binding EGF (HB-EGF) [66, 67]. HB-EGF interacts with both ErbB1 and ErbB4, and is an important trophic factor in the developing CNS [14]. As HB-EGF induces cerebral neurogenesis, it will be interesting to determine whether GnRH-mediated HB-EGF release has a role in neuronal migration and organization of the hypothalamic GnRH neuronal network [55]. EGF receptor ligands effect N-methyl-D-aspartate (NMDA)-receptor-mediated synaptic transmission [25, 48]. A similar mechanism of transactivation of ErbB1 and ErbB4 by a Gq-coupled, glutamate-mediated system through stimulation of metalloprotease activity has been described in astroglial cells in the hypothalamus [57, 70].

Transactivation of the FGF Receptor by GPCRs

Fibroblast growth factors (FGFs) are members of a family of polypeptides. At least four FGF receptors, FGFR1, FGFR2, FGFR3, and FGFR4, are expressed in the CNS [62]. FGF binding results in receptor dimerization and phosphorylation of specific tyrosine residues within the intracellular domain of the receptor. This, in turn, leads to activation of several intracellular signaling pathways, including a Ras/Raf/MEK/MAPK cascade. Moreover, FGF also activates other important signaling molecules such as PLCγ, Src, and PI3-K during cytoskeletal reorganization, differentiation, and survival in neuronal cells [47]. FGF-2 utilizes endogenous cannabinoid receptor 1 to promote axonal growth of cerebellar granule neurons [81]. Opioid receptors have been found to activate the FGF receptor and control nociception, stress responses, and addictive behaviors [38]. Activation of opioid receptors by morphine reduces naturally occurring neuronal death, while endogenous opioid peptides could act as trophic factors to regulate neuronal survival and cell proliferation during CNS development [38]. Stimulation of μ-opioid receptor in rat C6 glioma cells by morphine and endomorphine-1 caused a rapid increase in ERK1/2 phosphorylation through selective transactivation of FGFR1 [6]. Transactivation of FGFR1 by μ-opioid receptor involves Gβγ, Ca^{2+}, and PKC, as well as the metalloprotease-dependent release of FGF from glioma cells. This process is highly selective for the μ-opioid receptor, as indicated by the failure of κ-opioid receptors to induce FGFR1 phosphorylation. As the effects of the μ-opioid receptor are mediated through release of FGF, which in turn binds to its receptor, it is expected that μ-opioid receptor activation would mimic all actions of FGF. Also, because μ-opioid receptor activation prevents apoptosis following serum withdrawal in other cell types, including PC12 cells [21], differentiated SH-SY5Y cells, and cortical neurons [31], it will be of interest to see whether the effects of the μ-opioid receptor on these cells are mediated through transactivation of an RTK.

Transactivation of the PDGF Receptor by GPCRs

Platelet-derived growth factor (PDGF) is one of the most potent RTK mitogens and exerts its cellular effects by binding to and activating structurally related α and β receptors [22]. Whereas EGF receptor activation leads to enhanced synaptic transmission, activation of the PDGFβ receptor depresses glutamate receptor activity [41, 54]. D4 dopamine receptor activation in the CA1 region of the hippocampus caused depression of NMDA currents [36]. Because D4 dopamine receptors transactivate the PDGFβ receptor, the inhibition of PDGF receptor tyrosine kinase activity by selective inhibitors (WIN41662 and tyrphostin A9) blocked both PDGF- and quinpirole-induced depression of NMDA currents. Autophosphorylation of the tyrosine residues of PDGF receptors leads to activation of PLCγ, PI3-K, and Src [1]. Consistent with this, D4 dopamine receptor-mediated transactivation of PDGF receptors stimulated PLCγ-dependent release of intracellular Ca^{2+}, which enhanced Ca^{2+}-dependent inactivation of the NMDA receptor. This explains the mechanism of RTK-dependent synaptic transmission during GPCR activation in neuronal cells and implies a potential role of PDGF beyond cell survival and differentiation.

Transactivation of the Neurotrophin Receptor by GPCRs

Neurotrophins such as nerve growth factor (NGF), brain-derived neurotrophic factor (BDNF), neurotrophin 3 (NT3), and neurotrophin 4 (NT4) can signal through two different types of cell-surface receptor which have independent signaling properties: the tropomyosin-related kinase (Trk) receptor and the p75 neurotrophin receptor (p75NTR). NGF preferentially binds to trkA, BDNF and NT4 to trkB, and NT3 to trkC [64]. Neurotrophin signaling is an important factor in neuronal survival, proliferation, the fate of neural precursors, and axon and dendrite growth [15, 28]. Neurotrophins are synthesized as precursors that undergo maturation through proteolytic removal of the 'pro' region. Pro-neurotrophins and mature neurotrophins elicit opposite functional effects as a result of their different interactions with Trk and p75NTR receptors [15, 28]. Trk receptor activation regulates the expression and activity of functionally important proteins such as ion channels and neurotransmitter receptors, and it consequently regulates synaptic strength and plasticity [15]. By contrast, activation of p75NTR, most commonly by pro-neurotrophins, promotes cell death through activation of Rac and Jun N-terminal kinase (JNK) [15]. Tyrosine phosphorylation of the cytoplasmic domain of Trk receptors leads to the recruitment and activation of intracellular signaling molecules, including several small G proteins (Ras, Rap-1, Rac, and Rho), and signaling pathways dependent on MAPK, PI3-K, and PLCγ. The mechanism of GPCR-mediated Trk receptor activation has been described in the context of agonist stimulation by adenosine and pituitary adenylyl cyclase-activating polypeptides (PACAP). Adenosine acts as a neuromodulator and is upregulated when ATP is degraded and following hypoxic or ischemic injury [63]. PACAP is distributed throughout the CNS and has neurotrophic and neuroprotective effects in cerebellar granule neurons, motor neurons, mesencephalic dopaminergic neurons, basal forebrain cholinergic neurons, and cortical neurons, as well as in the peripheral nervous system [78]. The ability of adenosine and PACAP to mimic the effects of known neurotrophins results from adenosine-A2A receptor- or PACAP receptor-induced transactivation of trkA and trkB, as measured by tyrosine phosphorylation of the Trk receptors. Agonist-induced Trk transactivation depends on mobilization of intracellular Ca^{2+} and activation of Src tyrosine kinases [39, 40]. These studies provide evidence for the selective neurotrophic and neuroprotective effects of adenosine and could explain the observed similarities between the actions of GPCR agonists (adenosine and PACAP) and neurotrophins in the CNS. Whereas neurotrophin-induced neuronal survival depends on activation of both MAPK and PI3-K/Akt pathways, survival induced by adenosine and PACAP after neurotrophin (NGF and BDNF) withdrawal depends only on activation of the PI3-K/Akt system [39, 40]. Moreover, activation of ERK1/2 by adenosine and PACAP through Trk activation was much slower than that induced by other RTKs (e.g., EGF or PDGF receptors), possibly owing to the need for synthesis of new proteins. These studies show that GPCR agonists mimic the trophic effects of neurotrophins. Moreover, the effects of adenosine and PACAP on the development and plasticity of the neuronal circuitry [42, 63] closely resemble those of neurotrophins [15, 28, 44].

RTK Signaling Regulators in Neurons

The main function of RTK signaling is to modify the levels of proteins required for survival, axon growth, and guidance via changes in gene transcription, protein synthesis and degradation, and protein phosphorylation and dephosphorylation. RTK signaling during differentiation into neuronal lineages was analyzed using an Affymetrix microarray containing probes for over 30,000 genes. Pollard et al. [59] investigated the transcriptional profile of neural stem cells during FGF-mediated differentiation. They identified a set of 64 genes, including transcriptional regulators such as Mef2c, Nkx2.2, Olig1, Pou3, and Sox9. ERK1/2 and JNK pathways play a role in differentiation following NGF stimulation, regulating either the expression or phosphorylation of various target proteins, including c-Jun, Elk-1, activating transcription factor (ATF), neurofilament heavy chain (NF-H), and MAP kinase phosphatase (MKP) [77]. Elk-1 is a protein that belongs to the ETS domain superfamily of transcription factors and is typically known as a mitogen-activated transactivator involved in the regulation of proliferation in cells [71]. Stimulation of primary neural or neuroblastoma cells results in translocation of phosphorylated Elk-1 to the neuronal nucleus [20], which was shown to suppress the expression of pro-apoptotic genes in neuroblastoma cells [19]. Besides the regulation of proliferation, Elk-1 is likely to serve other functions in neurons, probably in the regulation of survival [24, 76] or neuronal apoptosis [3, 4]. Elk-1 mediates neuronal survival via the PI3-K/Akt pathway and was shown to rescue cells from apoptosis following serum withdrawal [13].

CREB is a transcription factor regulated by neurotrophins and ERK signaling. Elk-1 and CREB are phosphorylated in the dentate gyrus following LTP induction as a result of RTK activation [17]. Thus, different transcriptional regulators work in concert to regulate differentiation, migration, cell body positioning, axonal growth, and the survival of neurons during development. Some of them are also crucial in the adult neuron. CREB, for example, acts as an integration point for many GPCRs, Trk receptors, NMDA receptors, and L-type calcium channels and is thus implicated in a variety of different processes, including learning, memory, addiction, depression, anxiety, and many other higher cognitive functions, as well as in the plasticity associated with these processes [12]. CREB also regulates the expression of proteins such as c-Fos, BDNF, and glutamate receptor GluR1. Another target of MAPK signaling implicated in LTP, learning, and memory is the ribosomal S6 kinase (RSK2) [69]. RSK2 is a histone kinase [50]. *rsk2* gene mutations result in Coffin–Lowry syndrome (CLS), an X-linked disorder characterized by mild mental retardation and facial abnormalities [18].

Involvement of RTKs in the Pathophysiology of Neuropsychiatric Disease

Physiological Roles of PDGF and Dopamine Receptors in Neuropsychiatric Diseases

It has been reported that PDGF has neurotrophic effects on both GABAergic and dopaminergic neurons in the CNS and that PDGF has long-lasting effects on NMDA receptor-mediated synaptic transmission [41, 74]. Whereas EGF receptor activation leads to enhanced synaptic transmission, activation of PDGFβ receptors depresses NMDA receptor activity [36, 41, 74]. Dopamine, a major transmitter and neuromodulator in the CNS, is implicated in several psychiatric disorders, including schizophrenia, attention deficit hyperactivity disorder (ADHD), and Tourette's Syndrome [5]. Dopamine receptors have been classified into two categories based on sequence homology, signal transduction mechanisms, and agonist affinity: the D1 class receptors (the D1 and D5 receptors) and D2 class receptors (the D2, D3, and D4 receptors). D2 class dopamine receptors are thought to be the targets for most common antipsychotics [65]. Several cortical and limbic structures, and specifically the hippocampus, have been implicated in the deficits in cognition, memory, and attention associated with schizophrenia and ADHD [5]. Animal models have shown that D2 class receptors play an important role in attention and/or cognition, likely through modulation of the signaling mediated by the NMDA subtype of postsynaptic glutamate receptor [2, 32]. LTP in the hippocampus provides a model of learning and memory [7]. NMDA receptors are nonselective cation channels that are substantially permeable to Ca^{2+}, and the influx of Ca^{2+} through these channels is required to induce LTP [46, 48]. Paradoxically, the same Ca^{2+} influx that is required to induce LTP also acts via the Ca^{2+}-dependent regulatory protein calmodulin to inactivate these channels, a process referred to as Ca^{2+}-dependent inactivation. Calcium inactivation of NMDA receptors results in part from the displacement of the actin-binding protein α actinin 2 from a binding site on the NR1 subunit [37]. The hippocampal CA1 region is innervated by dopaminergic fibers [23] and expresses D2 and D4 receptors, but not D3 receptors [49, 53]. The selective activation of D4 dopamine receptors, both in acutely dissociated CA1 neurons and in the CA1 region of the hippocampus, but not in cultured neurons, results in the depression of NMDA currents. In heterologous cell systems, D4 dopamine receptors transactivate PDGFβ receptors [53] and, consistent with this observation, PDGF tyrosine kinase inhibitors (WIN41662 and tyrphostin A9), when added to the patch pipette, block both PDGF- and quinpirole-induced depression of NMDA currents. Furthermore, quinpirole treatment induces the internalization of PDGFβ receptors in acutely dissociated cultures and both PDGFβ receptor phosphorylation and MAPK activation in CA1 region hippocampal slices [36].

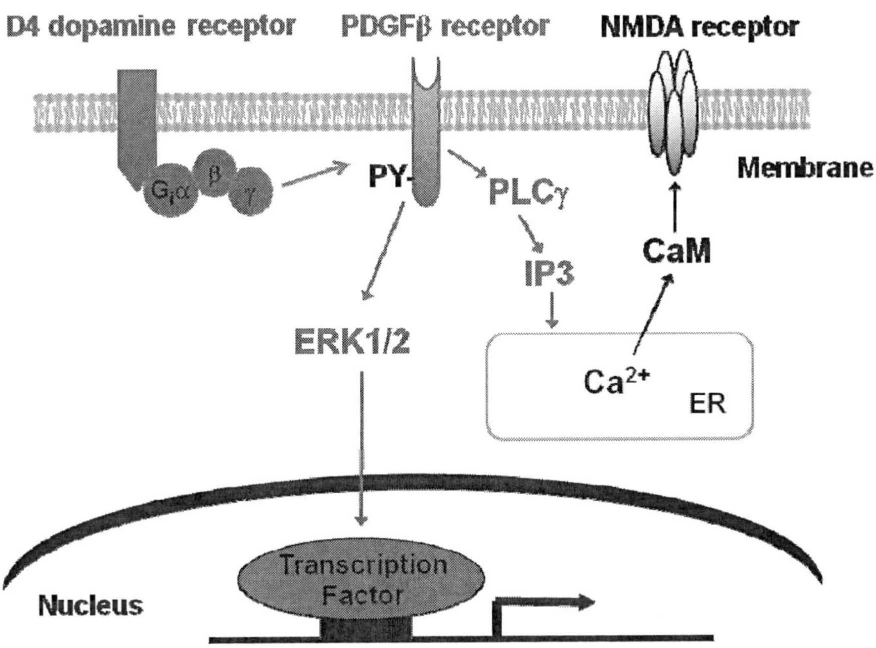

Figure 1. D4 dopamine receptor transactivation of the platelet-derived growth factor (PDGF) β receptor. Stimulation of D4 dopamine receptors causes autophosphorylation of the PDGFβ receptor on tyrosine residues (PY) through activation of Giα. PDGFβ receptor autophosphorylation leads to the recruitment of phospholipase Cγ (PLCγ). PLCγ-dependent increases in intracellular levels of inositol (1,4,5)-trisphosphate (IP3) results in the IP3 receptor-mediated release of Ca^{2+} from the endoplasmic reticulum (ER). The Ca^{2+}-dependent inactivation of NMDA currents is dependent on calmodulin (CaM). D4 dopamine receptor-mediated transactivation of the PDGFβ receptor also results in ERK1/2 activation and can affect the transcription of genes involved in cell growth, cell survival, and apoptosis in the central nervous system (CNS).

The autophosphorylation of tyrosine residues on the PDGFβ receptor provides intracellular docking sites for the recruitment and activation of effector enzymes such as PLCγ, PI3-K, and Src [26, 33]. Thus, D4 dopamine receptor transactivation of the PDGFβ receptor stimulates the PLCγ-dependent release of Ca^{2+} from intracellular stores, resulting in the inactivation of NMDA receptors, ERK1/2 activation, and phosphorylation of the MAPK substrate Elk-1 (Figure 1). The Ca^{2+}-dependent inactivation of NMDA currents is dependent on both calmodulin activity and the polymerization state of filamentous actin, but is independent of PI3-K, calmodulin kinase II (CaMKII), and calcineurin activity. These results indicate that dopamine receptors transactivate PDGF receptors and that RTK transactivation by GPCR has a pivotal role in synaptic transmission.

Novel Treatment Strategies for Neuropsychiatric Diseases

Dysfunction of dopamine receptor-mediated signaling has been implicated in neurological disorders that affect cognition and attention, such as schizophrenia and ADHD [5, 34]. The mainstay of schizophrenia treatment is antipsychotic medication, which primarily suppresses dopamine receptor activity [75]. On the other hand, the psychostimulant methylphenidate (MPH) is the first-choice treatment for ADHD. MPH inhibits the dopamine transporter (DAT), leading to increased dopamine levels in the brain [34]. Although reports have described the involvement of dopaminergic systems in the mechanism of action of psychotropic drugs, this mechanism of action is not completely understood. An abnormality in glutamate neurotransmission has also been proposed to be the basis of some forms of neuropsychiatric diseases [30]. Psychotropic drugs such as carbamazepine, imipramine, and related tricyclic antidepressants have been found to have antagonistic effects on NMDA receptors [27, 73]. There results indicate that both dopaminergic systems and NMDA receptor signaling are effective molecular targets in the treatment of neuropsychiatric disorders. Recent studies reported that PDGF receptors are associated with schizophrenia [10, 35] and that mice lacking the PDGFβ receptor display behavioral and neuroanatomical features that are typical of neuropsychiatric disorders [52]. These reports indicate that future studies of the physiological roles of PDGF receptor transactivation systems will also provide opportunities for producing novel treatment strategies for these diseases.

Despite the huge research efforts aimed at identifying high-affinity and specific agonists/antagonists with improved brain penetration for the treatment of CNS diseases, the achievements so far may seem somewhat disappointing. Problems of low efficacy and specificity for the molecular target contribute to an increased risk of adverse effects. Due to the lack of suitable assay systems, many projects have been aborted before they identified suitable therapeutic agents. Recently, induced pluripotent stem (iPS) cells were developed [72]. These cells can be generated directly from patients with CNS diseases and be directed to differentiate into neuronal cells [11]. Patient-specific iPS cells generated from individuals with CNS diseases would carry the precise constellation of genetic information associated with the pathology in that person. Many drug development trials for neuropsychiatric diseases have been aborted due to the lack of a suitable assay system. iPS-derived neurons would contribute to the establishment of a physiologically reliable assay system and screening of compound binding and disease-related RTK transactivation. Further opportunities for a breakthrough are offered by a deeper understanding of RTK transactivation, as this knowledge can be used to research drugs that have a more favorable interaction with signaling intermediates involved in a targeted pathological process. Drug combinations consisting of a GPCR and an RTK transactivation modifier or drugs with dual/multiple actions may also help us to find suitable therapeutic agents for the treatment of neuropsychiatric diseases.

Conclusions

GPCRs constitute the largest family of integral membrane proteins. They transduce information provided by extracellular signals (e.g., neurotransmitters) to the cell interior by interacting with heterotrimeric guanine nucleotide-binding proteins (i.e., G proteins). G proteins positively and negatively regulate the activity of effecter enzymes and ion channels. GPCR-stimulated second-messenger signals have a role in modulating synaptic activity and neurotransmission in the CNS. In addition to classical second-messenger-regulated mechanisms, GPCRs activate growth factor signaling cascades through the 'transactivation' of RTKs. The activation of dopamine receptors results in Gβγ subunit-dependent tyrosine autophosphorylation of the PDGFβ receptor. Dopamine receptor transactivation of the PDGFβ receptor leads to the inactivation of NMDA receptors and MAPK signaling. These signaling mechanisms have been implicated in neurological disorders. Physiological analysis of RTK transactivation in the CNS would contribute to a better understanding of neuropsychiatric diseases and the development of novel treatment strategies.

References

[1] Andrae J, Gallini R, Betsholtz C. Role of platelet-derived growth factors in physiology and medicine. *Genes Dev.* 2008. 22:1276-1312.

[2] Arnsten AF, Murphy B, Merchant K. The selective dopamine D4 receptor antagonist, PNU-101387G, prevents stress-induced cognitive deficits in monkeys. *Neuropsychopharmacology.* 2000. 23:405-410.

[3] Barrett LE, Sul JY, Takano H, Van Bockstaele EJ, Haydon PG, Eberwine JH. Region-directed phototransfection reveals the functional significance of a dendritically synthesized transcription factor. *Nat Methods.* 2006. 3:455-460.

[4] Barrett LE, Van Bockstaele EJ, Sul JY, Takano H, Haydon PG, Eberwine JH. Elk-1 associates with the mitochondrial permeability transition pore complex in neurons. *Proc Natl Acad Sci USA.* 2006. 103:5155-5160.

[5] Beaulieu JM, Gainetdinov RR. The physiology, signaling, and pharmacology of dopamine receptors. *Pharmacol Rev.* 2011. 63:182-217.

[6] Belcheva MM, Haas PD, Tan Y, Heaton VM, Coscia CJ. The fibroblast growth factor receptor is at the site of convergence between μ-opioid receptor and growth factor signaling pathways in rat C6 glioma cells. *J Pharmacol Exp Ther.* 2002. 303:909-918.

[7] Bliss TV, Collingridge GL. A synaptic model of memory: long-term potentiation in the hippocampus. *Nature.* 1993. 361:31-39.

[8] Bounanno A, Fischbach GD. Neuregulin and ErbB receptor signaling pathways in the nervous system. *Curr Opin Neurobiol.* 2001. 11:287-296.

[9] Bozon B, Kelly A, Josselyn SA, Silva AJ, Davis S, Laroche S. MAPK, CREB and zif268 are all required for the consolidation of recognition memory. *Philos Trans R Soc Lond B Biol Sci.* 2003. 358:805-814.

[10] Burbach JP, van der Zwaag B. Contact in the genetics of autism and schizophrenia. *Trends Neurosci.* 2009. 32:69-72.

[11] Chambers SM, Fasano CA, Papapetrou EP, Tomishima M, Sadelain M, Studer L. Highly efficient neural conversion of human ES and iPS cells by dual inhibition of SMAD signaling. *Nat Biotechnol.* 2009. 27:275-280.

[12] Carlezon WA, Jr, Duman RS, Nestler EJ. The many faces of CREB. *Trends Neurosci.* 2005. 28:436-445.

[13] Chang SH, Poser S, Xia Z. A novel role for serum response factor in neuronal survival. *J Neurosci.* 2004. 24:2277-2285.

[14] Chen Y, Fu AK, Ip NY. Bidirectional signaling of ErbB and Eph receptors at synapses. *Neuron Glia Biol.* 2008. 4:211-221.

[15] Chao MV. Neurotrophins and their receptors: a convergent point for many signaling pathways. *Nat Rev Neurosci.* 2003. 4:299-309.

[16] Cohen-Cory S. The developing synapse: construction and modulation of synaptic structures and circuits. *Science.* 2002. 298:770-776.

[17] Davies S, Vanhoutte P, Pages C, Caboche J, Laroche S. The MAPK/ERK cascade targets both Elk-1 and cAMP response element-binding protein to control long-term potentiation-dependent gene expression in the dentate gyrus in vivo. *J Neurosci.* 2000. 20:4563-4572.

[18] Delaunoy J, Abidi F, Zeniou M, Jacquot S, Merienne K, Pannetier S, Schmitt M, Schwartz C, Hanauer A. Mutations in the X-linked RSK2 gene (RPS6KA3) in patients with Coffin-Lowry syndrome. *Hum Mut.* 2001. 17:103-116.

[19] Demir O, Aksan Kurnaz I. Wildtype Elk-1, but not a SUMOylation mutant, represses egr-1 expression in SH-SY5Y neuroblastomas. *Neurosci Lett.* 2008. 437:20-24.

[20] Demir O, Korulu S, Yildiz A, Karabay A, Aksan Kurnaz I. Elk-1 interacts with neuronal microtubules and relocalizes to the nucleus upon phosphorylation. *Mol Cell Neurosci.* 2009. 40:111-119.

[21] Dermitzaki E, Chatzaki E, Gravanis A, Margioris AN. Opioids transiently prevent activation of apoptotic mechanisms following short periods of serum withdrawal. *J Neurochem.* 2000. 74:960-969.

[22] Fredriksson L, Li H, Eriksson U. The PDGF family: four gene products form five dimeric isoforms. *Cytokine Growth Factor Rev.* 2004. 15:197-204.

[23] Gasbarri A, Verney C, Innocenzi R, Campana E, Pacitti C. Mesolimbic dopaminergic neurons innervating the hippocampal formation in the rat: a combined retrograde tracing and immunohistochemical study. *Brain Res.* 1994. 994:71-79.

[24] Ginty D, Segal RA. Retrograde neurotrophin signaling: Trk-ing along the axon. *Curr Opin Neurobiol.* 2002. 12:268-274.

[25] Gu L, Li B, Yang X, Hu X, Huang X, Hertz L, Peng L. Depolarization-induced, glutamate receptor-mediated, and transactivation-dependent extracellular-signal regulated kinase phosphorylation in cultured cerebellar granule neurons. *Neuroscience.* 2007. 147:342-353.

[26] Heldin CH, Westermark B. Mechanism of action and in vivo role of platelet-derived growth factor. *Physiol Rev.* 1999. 79:1283-1316.

[27] Hough CJ, Irwin RP, Gao XM, Rogawski MA, Chuang DM. Carbamazepine inhibition of N-methyl-D-aspartate-evoked calcium influx in rat cerebellar granule cells. *J Pharmacol Exp Ther.* 1996. 276:143–149

[28] Huang EJ, Reichardt LF. TRK receptors: roles in neuronal signal transduction. *Annu Rev Biochem.* 2003. 72:609-642.

[29] Hubbard SR, Miller WT. Receptor tyrosine kinases: mechanisms of activation and signaling. *Curr Opin Cell Biol.* 2007. 19:117-123.

[30] Hui C, Wardwell B, Tsai GE. Novel therapies for schizophrenia: understanding the glutamatergic synapse and potential targets for altering N-methyl-D-aspartate neurotransmission. *Recent Pat CNS Drug Discov.* 2009. 4:220-238.

[31] Iglesias M, Segura MF, Comella JX, Olmos G. Mu-opioid receptor activation prevents apoptosis following serum withdrawal in differentiated SH-SY5Y cells and cortical neurons via phosphatidylinositol 3-kinase. *Neuropharmacology.* 2003. 44:482-492.

[32] Jentsch JD, Roth RH. The neuropsychopharmacology of phencyclidine: from NMDA receptor hypofunction to the dopamine hypothesis of schizophrenia. *Neuropsychopharmacology.* 1999. 20:201-225.

[33] Jurek A, Heldin CH, Lennartsson J. Platelet-derived growth factor-induced signaling pathways interconnect to regulate the temporal pattern of Erk1/2 phosphorylation. *Cell Signal.* 2011. 23:280-287.

[34] Kieling C, Genro JP, Hutz MH, Rohde LA. A current update on ADHD pharmacogenomics. *Pharmacogenomics.* 2010. 11:407-419.

[35] Kim HJ, Kim MH, Choe BK, Kim JW, Park JK, Cho AR, Bae H, Shin DH, Yim SV, Kwack K, Kwon YK, Chung JH. Genetic association between 5'-upstream single-nucleotide polymorphisms of PDGFRB and schizophrenia in a Korean population. *Schizophr Res.* 2008. 103:201-208.

[36] Kotecha SA, Oak JN, Jackson MF, Perez Y, Orser BA, Van Tol HH, MacDonald JF. D2 class dopamine receptor transactivates a receptor tyrosine kinase to inhibit NMDA receptor transmission. *Neuron.* 2002. 35:1111-1122.

[37] Krupp J, Vissel B, Thomas CG, Heinemann SF, Westbrook GL. Interactions of calmodulin and alpha-actinin with the NR1 subunit modulate Ca2+-dependent inactivation of NMDA receptors. *J Neurosci.* 1999. 99:165-1178.

[38] Law P-Y, Wong YH, Loh HH. Molecular mechanisms and regulation of opioid receptor signaling. *Annu Rev Pharmacol Toxicol.* 2000. 40:389-430.

[39] Lee FS, Chao MV. Activation of Trk neurotrophin receptors in the absence of neurotrophins. *Proc Natl Acad Sci USA.* 2001. 98:3555-3560.

[40] Lee FS, Rajagopal R, Kim AH, Chang PC, Chao MV. Activation of Trk neurotrophin receptor signaling by pituitary adenylate cyclase-activating polypeptides. *J Biol Chem.* 2002. 277:9096-9102.

[41] Lei S, Lu WY, Xiong ZG, Orser BA, Valenzuela CF, MacDonald JF. Platelet-derived growth factor receptor-induced feed-forward inhibition of excitatory transmission between hippocampal pyramidal neurons. *J Biol Chem. 1999.* 274:30617-30623.

[42] Liu Z, Geng L, Li R, He X, Zheng JQ, Xie Z. Frequency modulation of synchronized Ca2+ spikes in cultured hippocampal networks through G-protein-coupled receptors. *J Neurosci.* 2003. 23:4156-4163.

[43] Lowes VL, Ip NY, Wong YH. Integration of signals from receptor tyrosine kinases and g protein-coupled receptors. *Neurosignals.* 2002. 11:5-19.

[44] Lu B. BDNF and activity-dependent synaptic modulation. *Learn Mem.* 2003. 10:86-98.

[45] Luikart BW, Parada LF. Receptor tyrosine kinase B-mediated excitatory synaptogenesis. *Prog Brain Res.* 2006. 157:15-24.

[46] Malenka RC, Nicoll RA. Long-term potentiation-a decade of progress?, *Science.* 1999. 285:1870-1874.

[47] Mason I. Initiation to end point: the multiple roles of fibroblast growth factors in neural development. *Nat Rev Neurosci.* 2007. 8:583-596.

[48] McBain CJ, Mayer ML. N-methyl-D-aspartic acid receptor structure and function. *Physiol Rev.* 1994. 74:723-760.

[49] Meador-Woodruff JH, Grandy DK, Van Tol HH, Damask SP, Little KY, Civelli O, Watson SJ Jr. Dopamine receptor gene expression in the human medial temporal lobe. *Neuropsychopharmacology.* 1994. 10:239-248.

[50] Merienne K, Pannetier S, Harel-Bellan A, Sassone-Corsi P. Mitogen-regulated RSK2-CBP interaction controls their kinase and acetylase activities. *Mol Cell Biol.* 2001. 21:7089-7096.

[51] Naor Z. Signaling by G-protein-coupled receptor (GPCR): studies on the GnRH receptor. *Front Neuroendocrinol.* 2009. 30:10-29.

[52] Nguyen PT, Nakamura T, Hori E, Urakawa S, Uwano T, Zhao J, Li R, Bac ND, Hamashima T, Ishii Y, Matsushima T, Ono T, Sasahara M, Nishijo H. Cognitive and Socio-Emotional Deficits in Platelet-Derived Growth Factor Receptor-β Gene Knockout Mice. *PLoS One.* 2011. 6:e18004.

[53] Oak JN, Oldenhof J, Van Tol HH. The dopamine D4 receptor: one decade of research. *Eur J Pharmacol.* 2000. 405:303-327.

[54] Ohi Y, Ishii Y, Haji A, Noguchi S, Sasaoka T, Fujimori T, Nabeshima Y, Sasahara M, Hattori Y. Platelet-derived growth factor (PDGF)-BB inhibits AMPA receptor-mediated synaptic transmission via PDGF receptor-beta in murine nucleus tractus solitarius. *Brain Res.* 2007. 1159:77-85.

[55] Oyagi A, Oida Y, Kakefuda K, Shimazawa M, Shioda N, Moriguchi S, Kitaichi K, Nanba D, Yamaguchi K, Furuta Y, Fukunaga K, Higashiyama S, Hara H. Generation and characterization of conditional heparin-binding EGF-like growth factor knockout mice. *PLoS One.* 2009. 4:e7461.

[56] Pawson T, Gish GD, Nash P. SH2 domains, interaction modules and cellular wiring. *Trends Cell Biol.* 2001. 11:504-511.

[57] Peavy RD, Chang MS, Sanders-Bush E, Conn PJ. Metabotropic glutamate receptor 5-induced phosphorylation of extracellular signal-regulated kinase in astrocytes depends on transactivation of the epidermal growth factor receptors. *J Neurosci.* 2001. 21:9619-9628.

[58] Pierce KL, Premont RT, Lefkowitz RJ. Seven-transmembrane receptors. *Nat Rev Mol Cell Biol.* 2002. 3:639-650.

[59] Pollard SM, Wallbank R, Tomlinson S, Grotewold L, Smith A. Fibroblast growth factor induces a neural stem cell phenotype in fetal forebrain progenitors and during ES cell differentiation. *Mol Cell Neurosci.* 2008. 38:393-403.

[60] Prevot V, Rio C, Cho GJ, Lomniczi A, Heger S, Neville CM, Rosenthal NA, Ojeda SR, Corfas G. Normal female sexual development requires neuregulin-erbB receptor signaling in hypothalamic astrocytes. *J Neurosci.* 2003. 23:230-239.

[61] Ramos JW. The regulation of extracellular signal-regulated kinase (ERK) in mammalian cells. *Int J Biochem Cell Biol.* 2008. 40:2707-2719.

[62] Reuss B, Von Bohlen und Halbach O. Fibroblast growth factors and their receptors in the central nervous system. *Cell Tissue Res.* 2003. 313:139-157.

[63] Ribeiro JA. Adenosine A2A receptor interactions with receptors for other neurotransmitters and neuromodulators. *Eur J Pharmacol.* 1999. 375:101-113.

[64] Schecterson LC, Bothwell M. Neurotrophin receptors: Old friends with new partners. *Dev Neurobiol.* 2010. 70:332-338

[65] Seeman P, Van Tol HH. Dopamine receptor pharmacology. Trends Pharmacol Sci. 1994. 15:264-70.

[66] Shah BH, Soh JW, Catt KJ. Dependence of gonadotropin-releasing hormone-induced neuronal MAPK signaling on epidermal growth factor receptor transactivation. *J Biol Chem.* 2003. 278:2866-2875.

[67] Shah BH, Farshori MP, Jambusaria A, Catt KJ. Roles of Src and EGF receptor transactivation in transient and sustained ERK1/2 responses to GnRH receptor activation. *J Biol Chem.* 2003. 278:19118-19126.

[68] Shah BH, Catt KJ. GPCR-mediated transactivation of RTKs in the CNS: mechanisms and consequences. *Trends Neurosci.* 2004. 27:48-53.

[69] Sharma SK, Carew TJ. The roles of MAPK cascades in synaptic plasticity and memory in Aplysia: facilitatory effects and inhibitory constraints. *Learn Mem.* 2004. 11:373-378.

[70] Sharif A, Duhem-Tonnelle V, Allet C, Baroncini M, Loyens A, Kerr-Conte J, Collier F, Blond S, Ojeda SR, Junier MP, Prevot V. Differential erbB signaling in astrocytes from the cerebral cortex and the hypothalamus of the human brain. *Glia.* 2009. 57:362-379.

[71] Sharrocks AD. The ETS-domain transcription factor family. *Nat Rev Mol Cell Biol.* 2001. 2:827-837.

[72] Takahashi K, Tanabe K, Ohnuki M, Narita M, Ichisaka T, Tomoda K, Yamanaka S. Induction of pluripotent stem cells from adult human fibroblasts by defined factors. *Cell.* 2007. 131:861-872

[73] Tohda M, Urushihara H, Nomura Y. Inhibitory effects of antidepressants on NMDA-induced currents in Xenopus oocytes injected with rat brain RNA. *Neurochem Int.* 1995. 26:53-58.

[74] Valenzuela CF, Xiong Z, MacDonald JF, Weiner JL, Frazier CJ, Dunwiddie TV, Kazlauskas A, Whiting PJ, Harris RA. Platelet-derived growth factor receptor induces a long-term inhibition of N-methyl-D-aspartate receptor function. *J Biol Chem.* 1996. 271:16151-16159.

[75] van Os J and Kapur S. Schizophrenia. *Lancet.* 2009. 374:6356-6345.

[76] Vickers ER, Kazsa A, Aksan Kurnaz I, Seifert A, Zeef L, O'Donnell A, Hayes A, Sharrocks AD. Ternary TCF-SRF complex-regulated gene activity is required for cellular proliferation and inhibition of apoptotic cell death. *Mol Cell Biol.* 2004. 24:10340-10351.

[77] Waetzig V, Herdegen T. The concerted signaling of ERK1/2 and JNKs is essential for PC12 neuritogenesis and converges at the level of target proteins. *Mol Cell Neurosci.* 2003. 24:238-249.

[78] Waschek JA. Multiple actions of pituitary adenylyl cyclase activating peptide in nervous system development and regeneration. *Dev Neurosci.* 2002. 24:14-23.

[79] Werry TD, Sexton PM, Christopoulos A. "Ins and outs" of seven-transmembrane receptor signalling to ERK. *Trends Endocrinol Metab.* 2005. 16:26-33

[80] Wetzker R, Böhmer FD. Transactivation joins multiple tracks to the ERK/MAPK cascade. *Nat Rev Mol Cell Biol.* 2003. 4:651-657.

[81] Williams EJ, Walsh FS, Doherty P. The FGF receptor uses the endocannabinoid signaling system to couple to an axonal growth response. *J Cell Biol.* 2003. 160:481-486.

[82] Xian W, Schwertfeger KL, Rosen JM. Distinct roles of fibroblast growth factor receptor 1 and 2 in regulating cell survival and epithelial-mesenchymal transition. *Mol Endocrinol.* 2007. 21:987-1000.

[83] Yin G, Yan C, Berk BC. Angiotensin II signaling pathways mediated by tyrosine kinases. *Int J Biochem Cell Biol.* 2003. 35:780-783.

In: Tyrosine and Aspartic Acid
Editors: J. E. Jones and D. M. Morano
ISBN: 978-1-62100-752-4

Chapter 9

FUNCTIONS OF BRAIN L-ASPARTATE AND ITS DERIVATIVES ON STRESS RESPONSES

***Shozo Tomonaga*[1], *Edi Erwan*[2], *D. Michael Denbow*[3] *and Mitsuhiro Furuse*[2,*]**

[1]Laboratory of Advanced Animal and Marine Bioresources, Faculty of Agriculture, Kyushu University, Fukuoka, Japan
[2]Laboratory of Regulation in Metabolism and Behavior, Faculty of Agriculture, Kyushu University, Fukuoka, Japan
[3]Department of Animal and Poultry Sciences, Virginia Tech, Blacksburg, VA, US

ABSTRACT

L-Aspartate (L-Asp) functions as a neurotransmitter to stimulate the N-methyl-D-aspartate (NMDA) receptor (NMDA-R), one of the ionotropic L-glutamate receptors, even though its binding capacity for NMDA-R is weaker than L-glutamate. The amino acid L-Asp, as well as its enantiomer D-aspartate (D-Asp), occurs in the central nervous system of various species including chickens, pigeons, rats, mice and humans. D-Asp, synthesized from L-Asp by aspartate racemase, can also directly stimulate the NMDA-R. Furthermore, D-Asp may indirectly stimulate NMDA-R because it is a substrate for endogenous NMDA. Central injection of L-Asp and L-asparagine derived from L-Asp via transamidation decreased social separation-induced stress responses. In addition, central injection of NMDA can attenuate the stress responses while some reports indicate that stimulation of the NMDA-R has negative impacts on stress responses. Central L-Asp may act on stress responses not only directly but also via its metabolites, and the effects may depend on the type of stressors and/or brain regions.

* Correspondence to: Mitsuhiro Furuse, Laboratory of Regulation in Metabolism and Behavior, Faculty of Agriculture Kyushu University, Fukuoka 812-8511, Japan, Tel/Fax: (81) (92) 642-2953, E-mail: furuse@brs.kyushu-u.ac.jp.

1. INTRODUCTION

L-Aspartate (L-Asp) is one of the non-essential amino acids, and it plays various roles not only as a constituent of protein but also as part of bioactive substances. The metabolic pathway linked to L-Asp in animal tissues is shown in Figure 1. L-Asp can be metabolized to oxaloacetate or adenylosuccinate, which can enter the TCA cycle. L-Asp also can be metabolized to L-argininosuccinate and thus contribute to activation of the urea cycle. On the other hand, L-aspagine (L-Asn) is synthesized from L-Asp by asparagine synthetase. In this chapter, we mainly focused on the other pathway, starting from the synthesis of D-aspartate (D-Asp) by aspartate racemase. D-Asp occurs in the central nervous system of various species including chickens, pigeons, rats, mice and humans (Hashimoto et al., 1993, 1995; Kera et al., 1996; Morikawa et al., 2001; Neidle & Dunlop, 1990).

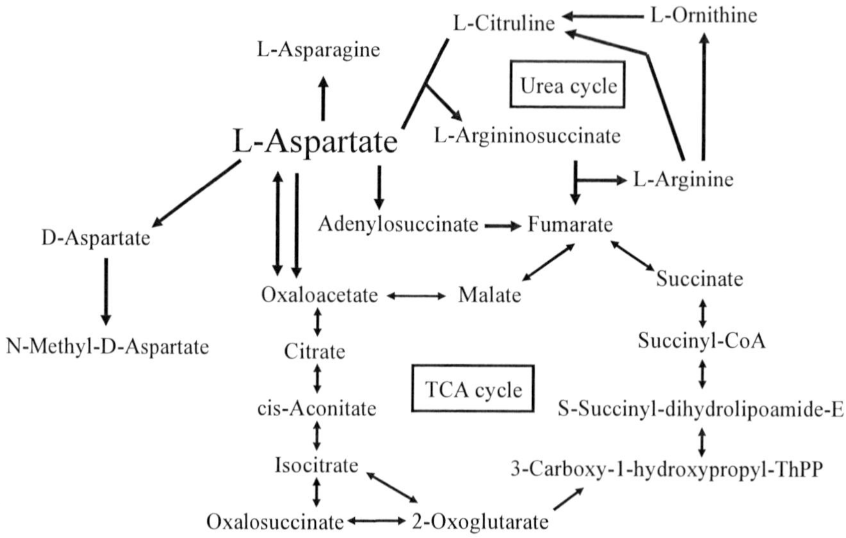

Figure 1. Metabolic pathways linked to L-aspartate.

Furthermore, D-Asp can be metabolized to N-methyl-D-aspartate (NMDA) (D'Aniello et al., 2000). L-Asp, D-Asp and NMDA can activate the NMDA-R in the brain (Kiskin et al., 1990). According to Hamasu et al. (2009), several amino acids including L-Asp and L-Asn were decreased in the diencephalon of neonatal chicks exposed to both restraint with isolation and fasting stress. This fact suggests that central free amino acids have an important role on the stress response. Furthermore, during stress the endocrine stress pathway involving the hypothalamic-pituitary-adrenal (HPA) axis is activated, and this activation is linked to excitatory neurotransmission involving the NMDA receptor (NMDA-R) (Zelena et al., 2005). However, the functions and mechanisms of this pathway have not been fully clarified.

Firstly, we discuss our study showing that L-Asp and L-Asn in the brain can attenuate acute stress responses in neonatal chicks (Yamane et al., 2009b). Then, we discuss the function of NMDA which had similar effects as observed with L-Asp (Yamane et al., 2009a) during the stress response. In these studies, the social separation stress model was used. This stress model is frequently used for the study of anxiety. Chicks are comfortable when living in a group, but exhibit anxiety when isolated. Social separation stress increases spontaneous activity and vocalization of chicks. This social separation stress paradigm has been used for developing anti-anxiety agents using vocalization and spontaneous activity as parameters (Feltenstein et al., 2003). Thirdly, we introduce the relationships between endocrine stress responses and the NMDA-R in the brain. Finally, possible roles of D-Asp for stress responses are discussed. The overall perspective of L-Asp and its metabolites for stress responses in the brain will be discussed.

2. Central Functions of L-Asp and L-Asn for Stress Behavior in Chicks

Glutamate and L-Asp function as neurotransmitters and are the most likely candidates for neurotransmitter action at excitatory amino acid receptors. There are two types of glutamate receptors, i.e., ionotropic and metabotropic. Agonists acting at ionotropic glutamate receptors increase the probability that the channel will open. There are three classes of glutamate ionotropic receptors (iGluRs), which are named after their selective agonists: NMDA, α-amino-3-hydroxy-5-methyl-4-isoxazole propionic acid (AMPA) and kainate (KA).

Of these receptors, L-Asp appears to recognize only the NMDA-R, being inactive at the AMPA, and probably the KA receptor (Dingledine & McBain 1999). The action of L-Asp on the NMDA-R was weaker than that of glutamate (Chen et al., 2005). We investigated the central effect of L-Asp, as well as L-Asn derived from L-Asp via transamidation, on the stress response.

Under acute social separation stress, the intracerebroventricular (i.c.v.) injection of L-Asp and L-Asn caused a sedative effect in neonatal chicks (Yamane et al., 2009b; Figure 2, Table 1). As described in the next section, i.c.v. injection of NMDA can also induce sedative effects. L-Asp acts at the NMDA-R (Dingledine & McBain 1999) and it acts as a selective NMDA-R agonist in avians (Kubrusly et al., 1998). Therefore, the effects of L-Asp may be mediated by the NMDA-R. Since L-Asn can be hydrolyzed by asparaginase to NH_4^+ and L-Asp, it is possible that L-Asn may have induced this effect either following chemical modification or by directly binding to the NMDA-R. This remains to be investigated.

L-Asp attenuated social-separation stress-induced behaviors during the 10 min post injection period, but enhanced plasma corticosterone concentration (Figure 3). Therefore, L-Asp might have little or no ability to suppress the secretion of corticosterone from the adrenal glands through the HPA axis in response to stress. According to Zhang et al. (2004), i.c.v. injection of serotonin increased plasma corticosterone. Their report suggested that serotonin stimulates the HPA axis. Furthermore, Zhang et al. (2004) reported that serotonin dose-dependently decreased locomotor activities of chicks following the central administration of 0.1 μg of corticotrophin-releasing factor (CRF). Kim et al. (1998) suggested that glutamatergic neurotransmission may modulate serotonergic function suggesting that the response to L-Asp may be mediated via a serotonergic pathway.

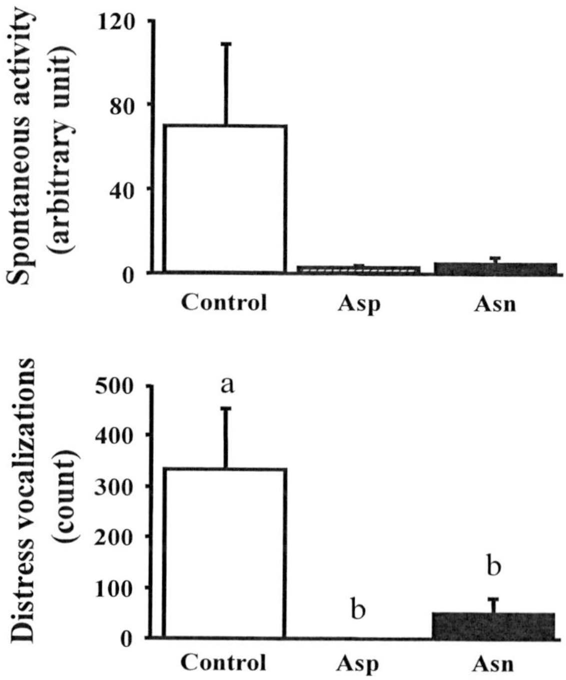

Figure 2. Effect of i.c.v. injection of saline, Asp and Asn on total spontaneous activity (upper panel) and distress vocalizations (lower panel) during a 10 min social-isolation period in 6-day-old layer chicks. Results are expressed as means ± SEM. The number of chicks used in each group was 9 in control, 7 in Asp and 8 in Asn groups. Groups with different letters are significantly different ($P<0.05$) Reproduced from Yamane, H., Asechi, M., Tsuneyoshi, Y., Kurauchi, I., Denbow, D.M., Furuse, M. (2009b) Intracerebroventricular injection of L-aspartic acid and L-asparagine induces sedative effects under an acute stressful condition in neonatal chicks. Animal Science Journal, 80:286-290, with permission from John Wiley & Sons.

Table 1. Effect of i.c.v. injection of saline, L-aspartic acid (Asp) and L-asparagine (Asn) on various behavioral categories of 6-day-old chicks during the 10 min post injection

	Control	Asp	Asn
Active wakefulness	$298{\pm}96^{a}$	$2{\pm}2^{b}$	$95{\pm}57^{ab}$
Standing/sitting motionless with eyes opened	67±34	129±35	145±50
Standing motionless with eyes closed	1±1	0±0	1±1
Sitting motionless with head drooped (sleeping posture)	234±93	469±37	359±72
Total	600	600	600

Values are means ± SEM in seconds. The number of chicks used in each group was 9 in control, 7 in Asp and 8 in Asn groups. Groups with different letters are significantly different ($P<0.05$). Reproduced from Yamane, H., Asechi, M., Tsuneyoshi, Y., Kurauchi, I., Denbow, D.M., Furuse, M. (2009b) Intracerebroventricular injection of L-aspartic acid and L-asparagine induces sedative effects under an acute stressful condition in neonatal chicks. Animal Science Journal, 80:286-290, with permission from John Wiley & Sons.

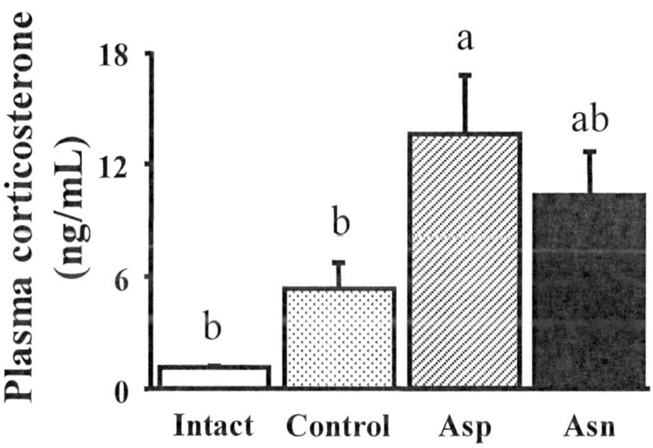

Figure 3. Effect of i.c.v. injection of saline, Asp and Asn on plasma corticosterone concentration following a 10 min social-isolation in 6-day-old layer chicks. Results are expressed as means ± SEM. The number of chicks used in each group was 5 in intact, 9 in control, 7 in Asp and 8 in Asn groups. Groups with different letters are significantly different ($P<0.05$). Reproduced from Yamane, H., Asechi, M., Tsuneyoshi, Y., Kurauchi, I., Denbow, D.M., Furuse, M. (2009b) Intracerebroventricular injection of L-aspartic acid and L-asparagine induces sedative effects under an acute stressful condition in neonatal chicks. Animal Science Journal. 2009 80:286-290, with permission from John Wiley & Sons

In conclusion, central L-Asp, as well as L-Asn, has a sedative effect under an acute stressful condition in neonatal chicks.

3. Central Functions of NMDA and Related Compounds for Stress Behavior in Chicks

Excitatory amino acids, including glutamate and L-Asp, act as neurotransmitters in the central nervous system. They can induce neuronal activity with powerful stimulatory effects (Monaghan et al., 1989). As stated above, there are three classes of iGluRs. The NMDA-R is usually blocked by Mg^{2+}. When the synaptic membrane is slightly depolarized, e.g., by previous activation of AMPA and KA receptors, the Mg^{2+} block of the NMDA-R is removed. The metabotropic glutamate receptors (mGluRs) are coupled to GTP-binding proteins, and regulate the production of intracellular messengers. The mGluRs have eight members (mGluR1-8), categorized into three groups based on sequence homology, second messenger coupling and pharmacology. Group I includes mGluR1 and 5, group II is mGluR2 and 3 and group III is mGluR4, 6, 7 and 8. Group I mGluRs are localized postsynaptically and predominantly activate phospholipase C. Group II and III mGluRs are localized in presynaptic densities and inhibit adenylyl cyclase activity. They contribute to the regulation of synaptic plasticity and transmission (De Blasi et al., 2001; Kew & Kemp, 2005).

The i.c.v. injection of NMDA, glutamate and AMPA attenuated total distress vocalizations and induced sedation in neonatal chicks (Yamane et al., 2009a: Figure 4, Table 2).

Table 2. Effect of i.c.v. injection of several doses of glutamate, N-methyhl-D-aspartate (NMDA), α-amino-3-hydroxy-5-methyl-4-isoxazole propionic acid (AMPA), kainate and 5-dehydroxyphenylglycine (DHPG) on various behavioral categories of chicks during 10 min post-injection

Drugs	Active wakefulness	Standing/sitting motionless with eyes open	Standing motionless with eyes closed	Sitting motionless with head drooped (sleeping posture)
L-Glutamate				
0 μmol	575±15	25±15	0±0	0±0
0.2 μmol	382±85*	120±39	11±11	87±53
0.4 μmol	159±46*	205±44*	29±29	207±69*
0.8 μmol	19±15*	134±23	10±10	437±26*
NMDA				
0 nmol	560±13	40±13	0±0	0±0
0.5 nmol	490±54	110±54	0±0	0±0
1 nmol	238±57*	262±41*	0±0	100±52
2 nmol	134±46*	314±53*	0±0	152±71
AMPA				
0 pmol	551±16	49±16	0±0	0±0
50 pmol	568±23	32±23	0±0	0±0
100 pmol	471±49	129±49	0±0	0±0
150 pmol	281±101*	102±34	10±7	207±79*
Kainate				
0 nmol	468±54	75±22	21±21	43±43
0.25 nmol	404±61	86±32	19±12	91±43
0.5 nmol	454±62	68±24	19±19	60±40
1 nmol	558±20	31±13	0±0	11±11
DHPG				
0 nmol	591±4	9±4	0±0	0±0
0.01 nmol	585±4	15±4	0±0	0±0
0.1 nmol	589±9	11±9	0±0	0±0
1 nmol	593±4	7±4	0±0	0±0

Values are means ± SEM in seconds. *Significantly different when compared with control at $P < 0.05$. L-Glutamate; Active wakefulness (second/10min) = 526 (SE= 41) – 692 (SE= 89) X (R^2= 0.701, $P < 0.0001$) and sleeping posture (second/10 min) = -11 (SE= 34) + 553 (SE= 74) X (R^2= 0.684, $P < 0.0001$).

NMDA; Active wakefulness (second/10 min) = 552 (SE= 38) - 223 (SE= 32) X (R^2= 0.640, $P < 0.0001$) and standing/sitting motionless with eyes open (second/10min) = 58 (SE = 35) + 140 (SE = 30) X (R^2 = 0.450, $P < 0.0001$).

AMPA; Active wakefulness (second/10min) = 600 (SE= 51) – 2 (SE= 1) X (R^2= 0.309, $P < 0.01$) and sleeping posture (second/10 min) = -40 (SE= 38) + 1 (SE= 0.4) X (R^2= 0.274, $P < 0.01$). Reproduced from Yamane, H., Tsuneyoshi, Y., Denbow, D.M., Furuse, M. (2009a) N-Methyl-D-aspartate and α-amino-3-hydroxy-5-methyl-4-isoxazolepropionate receptors involved in the induction of sedative effects under an acute stress in neonatal chicks. Amino Acids, 37:733-739.Erratum, 37:767, with permission from Springer-Verlag.

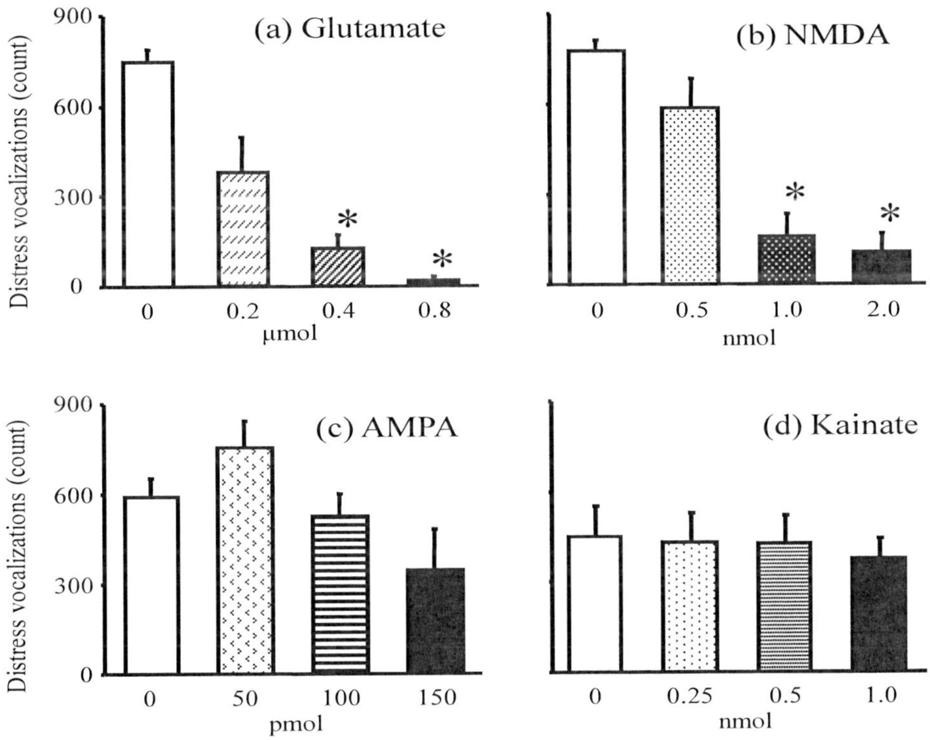

Figure 4. Effect of i.c.v. injection of iGluR agonists on total distress vocalizations during a 10 min isolation period in chicks. Results are expressed as means with S.E.M. Glutamate (count/10 min) = 621 (SE = 62) – 868 (SE = 135) X (R^2= 0.613, $P < 0.0001$). NMDA (count/10 min) = 708 (SE = 64) – 342 (SE = 54) X (R^2= 0.596, $P < 0.0001$). AMPA (count/10 min) = 691 (SE = 83) – 2 (SE = 1) X (R^2= 0.158, $P < 0.05$). *Significant difference when compared with the control at $P < 0.05$. Reproduced from Yamane, H., Tsuneyoshi, Y., Denbow, D.M., Furuse, M. (2009a) N-Methyl-D-aspartate and α-amino-3-hydroxy-5-methyl-4-isoxazolepropionate receptors involved in the induction of sedative effects under an acute stress in neonatal chicks. Amino Acids, 37:733-739. Erratum, 37:767, with permission from Springer-Verlag.

In contrast, previous studies showed that localized infusion of glutamate induced vocalizations in cats (Bandler, 1982) and squirrel monkeys (Jürgens & Richter, 1986). Additionally, the injection of NMDA, AMPA and KA into the substantia innominata/lateral preoptic area of rats dose-dependently stimulated locomotion (Shreve & Uretsky, 1988). Lateral hypothalamic injection of KA also increased locomotor activity in rats (Stanley et al., 1993; Hettes et al., 2007). Further, when injected intracranially to restricted brainstem regions, NMDA elicited locomotion in decerebrate geese and ducks (Sholomenko et al., 1991). On the other hand, injection of glutamate into the lateral hypothalamus increased sleeping time (Stanley et al., 1993). The difference between these studies and Yamane et al. (2009a) may be due to species differences or the site of injection. The i.c.v. injection of glutamate dose-dependently decreased distress vocalizations in isolated chicks similarly to that observed by Panksepp et al. (1988) who injected glutamate at doses from 25 to 500 μg. In contrast, Panksepp et al. (1988) reported the administration of NMDA (0.1, 0.25, 0.5 and 1.0 μg) had no effect on the number of vocalizations, which is inconsistent with our results.

Differences in the time for behavioral observation, experimental conditions, or genetic lines of chicks may contribute to the contrasting results for vocalizations. For example, Saito et al. (2005) found the stress response differs between meat- and layer-type neonatal chicks with this response being relatively higher in the layer-type used by Yamane et al. (2009a).

Although i.c.v. injection of glutamate, and the glutamate analogue NMDA, induced sedation in chicks, the behavioral results for the two compounds were somewhat different. NMDA induced an increase in the time spent standing/sitting motionless with eyes open while glutamate increased the time of sleeping posture. AMPA had a tendency to decrease wakefulness activity and increase sleep-like behavior. The central administration of KA (0.05, 0.1, 0.25 and 0.5 μg) had no effect under stressful condition in chicks (Panksepp et al., 1988). No significant effect of KA was also observed in our laboratory (Yamane et al., 2009a).

Henley et al. (1989) demonstrated the distribution of iGluRs ligand binding sites in the chick brain using autoradiography. The binding sites of NMDA and AMPA are related to the brain regions of stress response. However, KA binding sites are mainly the molecular layer of the cerebellum which is not the stress related region in the brain. In general, the NMDA-R usually coexists with the AMPA receptor in the postsynaptic membrane (Nadler, 2007). Therefore, KA may not be involved in the stress response. In any case, KA, in the doses used here, appeared to have no sedative effect against isolation stress in chicks.

It is suggested that the hypnotic effects of glutamate involved not only the NMDA-R, but also the AMPA receptor. Yamane et al. (2009a) applied the group I selective mGluRs agonist 5-dehydroxyphenylglycine (DHPG), but no effect on sleeping posture was detected. Although we did not investigate the effect of group II and III mGluRs, the data supports the theory that the iGluRs NMDA and AMPA, but not KA, are related to sleep-induction in neonatal chicks. Glutamate can bind to both AMPA and KA receptors; its action is probably mediated via NMDA, AMPA and KA receptors. On the other hand, the group I selective mGluRs agonist DHPG had no effect. Thus, it appears likely that the hypnotic and sedative effects of glutamate are induced by interaction with NMDA and AMPA receptors. However, the activity of group II mGluRs regulates the HPA axis (Scaccianoce et al., 2003), and group III mGluRs agonists have anxiolytic- and antidepressant-like effects in rats (Pałucha et al., 2004). Therefore, it is necessary to further investigate the interaction between mGluRs and stress behavior.

In conclusion, the central administration of glutamate attenuates stress-induced behaviors and triggers sleep-like behavior in neonatal chicks. Furthermore, it was demonstrated that NMDA and AMPA, but not KA and DHPG, also have a sedative effect. It is suggested that glutamate induced hypnosis and sedation may be necessary for the interaction between NMDA and AMPA receptors. Therefore, studies involving the co-administration with NMDA or AMPA antagonists are needed to better understand the mechanism of glutamate.

4. Functions of Brain NMDA Receptor on Endocrine Stress Axis

In this section, we briefly reviewed the functions of the NMDA-R on the endocrine stress axis because L-Asp, D-Asp and NMDA can all stimulate the NMDA-R (Kiskin et al., 1990). The HPA axis is a main endocrine stress axis responding to a diverse set of stressful stimuli.

CRF from the paraventricular nucleus (PVN) of the hypothalamus stimulates adrenocorticotropic hormone (ACTH) release from the anterior pituitary into the blood. Then, ACTH goes to the adrenal cortex and stimulates release of glucocorticoids. Anatomical, pharmacological, and electrophysiological studies indicate that glutamate is a major excitatory signal to the HPA axis (Herman et al., 2001). Focusing on the NMDA-R, acute immobilization stress increases NMDA-R1, one of the NMDA-R subunits, mRNA in the hypothalamic PVN (Bartanusz et al., 1995). In contrast, osmotic stress induced by water deprivation does not increase NMDA-R expression in the PVN (Meeker et al., 1994), suggesting that acute psychological stress may have a more profound affect on activating glutamatergic input to the PVN. On the other hand, chronic intermittent stress induces a significant decrease in the NR2B subunit of NMDA-R mRNA expression in the CRF-containing region of the PVN in rats (Ziegler et al., 2005). Zelena et al. (2005) suggested that intravenously administered NMDA (5 mg/kg) elevated ACTH and glucocorticoid levels while PVN lesions had no influence on their basal secretion but blocked NMDA-induced elevations. Therefore, NMDA may activate the HPA axis at a central (PVN) level and not at the level of pituitary or adrenal gland in rats.

5. Possible Functions of Central D-Asp for Stress Responses

D-Amino acids, the enantiomers of L-amino acids, are candidates to be novel physiologically active substances and biomarkers in animals, but D-amino acids demonstrate different actions and mechanisms in the central nervous system. For instance, L-serine might be converted to the enantiomer D-serine by serine racemase, which has been purified from the mammalian brain (Wolosker et al., 1999). D-Serine is present in the brain of several vertebrate species including carp, frog, mice, rat and chick (Nagata et al., 1994), and is in particularly high concentration in the mammalian brain where it acts as an endogenous ligand for the NMDA-R-related glycine site. According to Asechi et al. (2006), L-serine induced sedative and hypnotic effects in chicks under acute stress, but D-serine did not. The ligand for the NMDA-R, glycine, had similar sedative and hypnotic effects under acute stress. Why D-serine did not show sedative and hypnotic effects was unclear, but glycine can act at the glycine receptor. L-Serine is a precursor for the synthesis of other amino acids including glycine and L-cysteine. Asechi et al. (2006) confirmed that L-cysteine, as well as L-serine and glycine, had sedative and hypnotic effects in chicks under acute stress. However, L- and D-cysteine had a different mechanism under acute stress since L-cysteine increased sleep-like behavior while D-cysteine caused abnormal behavior including syncope as well as sleep-like behavior (Yamane et al., 2009c). On the other hand, it was an interesting observation that L- and D-proline functioned through different receptors. L-Proline and D-proline differentially induce sedative and hypnotic effects through NMDA and glycine receptors, respectively (Hamasu et al., 2010).

D-Asp has been detected in various proteins from human tissue samples such as eye lens, brain, skin, bone, teeth and aorta from elderly individuals (Fujii, 2005). The presence of D-Asp is believed to be the result of racemization of L-Asp. However, a chiral reaction field exists in the native higher order structure of protein which induces the inversion of L-Asp to D-Asp residue (Fujii, 2005). Free D-Asp occurs in the central nervous system of various species including chickens, pigeons, rats, mice and humans (Hashimoto et al., 1993, 1995; Kera et al., 1996; Morikawa et al., 2001; Neidle & Dunlop, 1990). To our knowledge, there are few reports investigating the effects of D-Asp on stress responses. From discussions in the previous sections, it can be easily hypothesized that brain D-Asp has sedative effects under social isolation stress in chicks and it can stimulate the HPA-axis by acting on the NMDA-R not only by itself but also by its metabolite NMDA. On the other hand, recently Molinaro et al. (2010) suggests that D-Asp, but not NMDA, activates mGluR5 receptors coupled to polyphosphoinositide hydrolysis in early postnatal rat brain slices. mGluR5 is highly expressed during the early postnatal brain and is also found in the embryonic brain (Di Giorgi Gerevini et al., 2004). mGluR5 could play some role in the stress response because stress-induced hyperthermia was reduced in mGluR5 knockout mice (Brodkin et al., 2002). Therefore, behavioral and/or endocrine stress responses may be regulated by brain D-Asp to stimulate both NMDA-R and mGluR5 especially in the developmental period. Stress in this period could induce a big impact (both negatively and positively) on brain maturation and/or function such as stress sensitivity in adulthood (Parker et al., 2006; Kikusui & Mori, 2009). Therefore, clarification of the relationship between central D-Asp and stress response in this period may be important to understand the physiological mechanisms of stress response in the brain. This should be investigated in the future.

Acknowledgments

A part of studies described here was supported by grant-in-aid for scientific research from Japan Society for the Promotion of Science (Nos. 17208023, 18208023 and 23248046) and Uehara Memorial Fundation to MF. This work was also supported by a Research Fellowship of the Japan Society for the Promotion of Science for Young Scientist to ST (No. 189592 and 22780240).

References

Asechi, M., Tomonaga, S., Tachibana, T., Han, L., Hayamizu, K., Denbow, D.M., Furuse, M. (2006) Intracerebroventricular injection of L-serine analogs and derivatives induces sedative and hypnotic effects under an acute stressful condition in neonatal chicks. *Behavioural Brain Research,* 170:71-77.

Bandler, R. (1982) Induction of 'rage' following microinjections of glutamate into midbrain but not hypothalamus of cat. *Neuroscience Letters,* 30:183-188.

Bartanusz, V., Aubry, J.M., Pagliusi, S., Jezova, D., Baffi, J., Kiss, J.Z. (1995) Stress-induced changes in messenger RNA levels of N-methyl-Daspartate and AMPA receptor subunits in selected regions of the rat hippocampus and hypothalamus. *Neuroscience,* 66:247–252.

Brodkin, J., Bradbury, M., Busse, C., Warren, N., Bristow, L.J., Varney, M.A. (2002) Reduced stress-induced hyperthermia in mGluR5 knockout mice. *European Journal of Neuroscience,* 16:2241-2244.

Chen, P.E., Geballe, M.T., Stansfeld, P.J., Johnston, A.R., Yuan, H., Jacob, A.L., Snyder, J.P., Traynelis, S.F., Wyllie, D.J.A. (2005) Structural features of the glutamate binding site in recombinant NR1/NR2A N-methyl-D-aspartate receptors determined by site-directed mutagenesis and molecular modeling. *Molecular Pharmacology,* 67:1470-1484.

D'Aniello, G., Tolino, A., D'Aniello, A., Errico, F., Fisher, G.H., Di Fiore, M.M. (2000) The role of D-aspartic acid and N-methyl-D-aspartic acid in the regulation of prolactin release. *Endocrinology,* 141:3862-3870.

De Blasi, A., Conn, P.J., Pin, J.P., Nicoletti, F. (2001) Molecular determinants of metabotropic glutamate receptor signaling. *Trends in Pharmacological Sciences,* 22:114-120.

Di Giorgi Gerevini, V.D., Caruso, A., Cappuccio, I., Ricci Vitiani, L., Romeo, S., Della Rocca, C., Gradini, R., Melchiorri, D., Nicoletti, F. (2004) The mGlu5 metabotropic glutamate receptor is expressed in zones of active neurogenesis of the embryonic and postnatal brain. *Brain Research, Developmental Brain Research,* 150:17-22.

Dingledine, R., McBain, C.J. (1999) Glutamate and Aspartate. In: Basic Neurochemistry. Molecular, *Cellular and Medical Aspects.* (Siegel G., Agranoff B.W., Albers R.W., Fisher S.K. and Uhler M.D. eds.). Sixth Edition. pp. 315-333. Lippincott-Raven Publishers. New York.

Feltenstein, M.W., Lambdin, L.C., Ganzera, M., Ranjith, H., Dharmaratne, W., Nanayakkara, N.P., Khan, I.A., Sufka, K.J. (2003) Anxiolytic properties of Piper methysticum extract samples and fractions in the chick. *Phytotherapy Research,* 17:210-216.

Fujii, N. (2005) *D-amino Biological & Pharmaceutical Bulletin,* 28:1585-1589.

Hamasu, K., Haraguchi, T., Kabuki, Y., Adachi, N., Tomonaga, S., Sato, H., Denbow, D.M., Furuse, M. (2009) L-Proline is a sedative regulator of acute stress in the brain of neonatal chicks. *Amino Acids,* 37:377-382.

Hamasu, K., Shigemi, K., Tsuneyoshi, Y., Yamane, H., Sato, H., Denbow, D.M., Furuse, M. (2010) Intracerebroventricular injection of L-proline and D-proline induces sedative and hypnotic effects by different mechanisms under an acute stressful condition in chicks. *Amino Acids,* 38:57-64.

Hashimoto, A., Kumashiro, S., Nishikawa, T., Oka, T., Takahashi, K., Mito, T., Takashima, S., Doi, N., Mizutani, Y., Kaneco, T., Ootomo, E. (1993) Embryonic development and postnatal changes in free D-aspartate and D-serine in the human prefrontal cortex. *Journal of Neurochemistry,* 61:348–351.

Hashimoto, A., Oka, T., Nishikawa, T. (1995) Anatomical distribution and postnatal changes in endogenous free D-aspartate and D-serine in rat brain and periphery. *European Journal of Neuroscience,* 7:1657–1663.

Henley, J.M., Moratallo, R., Hunt, S.P., Barnard, E.A. (1989) Localization and quantitative autoradiography of glutamatergic ligand binding sites in chick brain. *European Journal of Nuroscience,* 1:516-523.

Herman, J.P., Tasker, J.G., Ziegler, D.R., Cullinan, W.E. (2001) Paraventricular hypothalamic stress integration: limbic connections with local circuit neurons. *Pharmacology Biochemistry and Behavior,* 71:457–468.

Hettes, S.R., Heyming, W., Stanley, B.G. (2007) Stimulation of lateral hypothalamic kainate receptors selectively elicits feeding behavior. *Brain Research,* 1184:178-185.

Jürgens, U., Richter, K. (1986) Glutamate-induced vocalization in the squirrel monkey. *Brain Research,* 373:349-358.

Kera, Y., Aoyama, H., Watanabe, N., Yamada R.H. (1996) Distribution of D-aspartate oxidase and free D-glutamate and D-aspartate in chicken and pigeon tissues.

Comparative Biochemistry and Physiology Part B: Biochemistry and Molecular Biology, 115:121-126.

Kew, J.N.C., Kemp, J.A. (2005) Ionotropic and metabotropic glutamate receptor structure and pharmacology. *Psychopharmacology* (Berl), 179:4-29.

Kikusui, T., Mori, Y. (2009) Behavioural and neurochemical consequences of early weaning in rodents. Journal of Neuroendocrinology, 21:427-431.

Kim, H.S., Park, I.S., Park, W.K. (1998) NMDA receptor antagonists enhance 5-HT2 receptor-mediated behavior, head-twitch response, in mice. *Life Sciences,* 63:2305-2311.

Kiskin, N.I., Krishtal, O.A., Tsyndrenko, A.Y. (1990) Cross-desensitization reveals pharmacological specificity of excitatory amino acid receptors in isolated hippocampal neurons. *European Journal of Neuroscience,* 2:461-470.

Kubrusly, R.C., de Mello, M.C., de Mello, F.G. (1998) Aspartate as a selective NMDA receptor agonist in cultured cells from the avian retina. *Neurochemistry International,* 32:47-52.

Meeker, R.B., McGinnis, S., Greenwood, R.S., Hayward, J.N. (1994) Increased hypothalamic glutamate receptors induced by water deprivation. *Neuroendocrinology,* 60:477-485.

Molinaro, G., Pietracupa, S., Di Menna, L., Pescatori, L., Usiello, A., Battaglia, G., Nicoletti, F., Bruno, V. (2010) D-aspartate activates mGlu receptors coupled to polyphosphoinositide hydrolysis in neonate rat brain slices. *Neuroscience Letters,* 478:128-130.

Monaghan, D.T., Bridges, R.J., Cotman, C.W. (1989) The excitatory amino acid receptors: Their classes, pharmacology, and distinct properties in the function of the central nervous system. *Annual Review of Pharmacology and Toxicology,* 29:365-402.

Morikawa, A., Hamase, H., Inoue, T., Konno, R., Niwa, A., Zaitsu, K. (2001) Determination of free D-aspartic acid, D-serine and D-alanine in the brain of mutant mice lacking D-amino acid oxidase activity. *Journal of Chromatography B: Biomedical Sciences and Applications,* 757:119–125.

Nadler, J.V. (2007) Encyclopedia of stress. In: Fink G, McEwen B, Ronald de Kloet E, Rubin R, Chrousos G, Steptoe A, Rose N, Craig I, Feuerstein G (ed) *Excitatory amino acids,* 2nd edn. Elsevier, Australia, pp. 970-975.

Nagata, Y., Horiike, K., Maeda, T. (1994) Distribution of free D-serine in vertebrate brains. *Brain Research,* 634: 291-295.

Neidle, A., Dunlop, D.S. (1990) Developmental changes of free D-aspartic acid in the chicken embryo and in the neonatal rat. *Life Sciences,* 46:1517–1522.

Pałucha, A., Tatarczyńska, E., Brański, P., Szewczyk, B., Wierońska, J.M., Kłak, K., Chojnacka-Wójcik, E., Nowak, G., Pilc, A. (2004) Group III mGlu receptor agonists produce anxiolytic- and antidepressant-like effects after central administration in rats. *Neuropharmacology,* 46:151-159

Panksepp, J., Normasell, L., Herman, B., Bishop, P., Crepeau, L. (1988) The physiological control of mammalian vocalization. In: Newman J.D. (ed) *Neural and neurochemical control of the separation distress call,* Plenum Press, New York, pp. 263-299.

Parker, K.J., Buckmaster, C.L., Sundlass, K., Schatzberg, A.F., Lyons, D.M. (2006) Maternal mediation, stress inoculation, and the development of neuroendocrine stress resistance in primates. *Proceedings of the National Academy of Sciences of the United States of America,* 103:3000-3005.

Saito, S., Tachibana, T., Choi, Y.H., Denbow, D.M., Furuse, M. (2005) ICV CRF and isolation stress differentially enhance plasma corticosterone concentrations in layer- and meat-type neonatal chicks. *Comparative Biochemistry and Physiology Part A: Molecular & Integrative Physiology*, 141:305-309.

Scaccianoce, S., Matrisciano, F., Del Bianco, P., Caricasole, A., Gerevini, V.D.G., Cappuccio, I., Melchiorri, D., Battaglia, G., Nicoletti, F. (2003) Endogenous activation of group-II metabotropic glutamate receptors inhibits the hypothalamic-pituitary-adrenocortical axis. *Neuropharmacology* (Berl), 44:555-561.

Sholomenko, G.N., Funk, G.D., Steeves, J.D. (1991) Avian locomotion activated by brainstem infusion of neurotransmitter agonists and antagonists. I. Acetylcholine, excitatory amino acids and substance P. *Experimental Brain Research,* 85:659-673.

Shreve, P.E., Uretsky, N.J. (1988) AMPA, kainic acid, and N-methyl-D-Aspartic acid stimulate locomotor activity after injection into the substantia innominata/lateral preoptic area. *Pharmacology Biochemistry and Behavior,* 34:101-106.

Stanley, B.G., Ha, L.H., Spears, L.C., Dee, M.G. II (1993) Lateral hypothalamic injections of glutamate, kainic acid, D, L-α-amino-3-hydroxy-5-methyl-isoxazole propionic acid or N-methyl-D-aspartic acid rapidly elicit intense transient eating in rats. *Brain Research,* 613:88-95.

Wolosker, H., Blackshaw, S., Snyder, S.H. (1999) Serine racemase: a glial enzyme synthesizing D-serine to regulate glutamate-N-methyl-D-aspartate neurotransmission. *Proceedings of the National Academy of Sciences of the United States of America,* 96:13409-13414.

Yamane, H., Tsuneyoshi, Y., Denbow, D.M., Furuse, M. (2009a) N-Methyl-D-aspartate and α-amino-3-hydroxy-5-methyl-4-isoxazolepropionate receptors involved in the induction of sedative effects under an acute stress in neonatal chicks. *Amino Acids,* 37:733-739.Erratum, 37:767.

Yamane, H., Asechi, M., Tsuneyoshi, Y., Kurauchi, I., Denbow, D.M., Furuse, M. (2009b) Intracerebroventricular injection of L-aspartic acid and L-asparagine induces sedative effects under an acute stressful condition in neonatal chicks. *Animal Science Journal,* 80:286-290.

Yamane, H., Asechi, M., Tsuneyoshi, Y., Denbow, D.M., Furuse, M. (2009c) Central L-cysteine induces sleep, and D-cysteine induces sleep and abnormal behavior during acute stress in neonatal chicks. *Animal Science Journal,* 80:428-432.

Zelena, D., Mergl, Z., Makara, G.B. (2005) Glutamate agonists activate the hypothalamic-pituitary-adrenal axis through hypothalamic paraventricular nucleus but not through vasopressinerg neurons. *Brain Research,* 1031:185-193.

Zhang, R., Tachibana, T., Takagi, T., Koutoku, T., Denbow, D.M., Furuse, M. (2004) Serotonin modifies corticotropin-releasing factor-induced behaviors of chicks. Behavioural Brain Research, 151:47-52.

Ziegler, D.R., Cullinan, W.E., Herman, J.P. (2005) Organization and regulation of paraventricular nucleus glutamate signaling systems: N-methyl-D-aspartate receptors. *Journal of Comparative Neurology*, 484:43-56.

Index

A

B

C

D

E

I

N

O

P

Q

R

S

T

U

V

W

X

Y